Max S. Power uses his depth of experience and knowledge to summarize the key historical issues related to nuclear waste management in the United States. And, he takes into account the many different perspectives that must be considered in long-term strategy.

—*John E. Till, Ph.D., President, Risk Assessment Corporation*

America's Nuclear Wastelands manages to capture 60 years of nuclear history, tackling political, societal, and technical facets in a readable and entertaining manner. Facts are presented in an objective and non-judgmental manner without diminishing the terrifying aspects of the crisis that faces our nation and future generations. The author's greatest accomplishment is in making complicated scientific and technical issues accessible to all readers.

—*Tom Carpenter, Executive Director, Hanford Challenge*

AMERICA'S NUCLEAR WASTELANDS

AMERICA'S NUCLEAR WASTELANDS

POLITICS, ACCOUNTABILITY, AND CLEANUP

Max S. Power

Washington State University Press
Pullman, Washington

Washington State University Press
PO Box 645910
Pullman, Washington 99164-5910
Phone: 800-354-7360
Fax: 509-335-8568
E-mail: wsupress@wsu.edu
Web site: wsupress.wsu.edu

Library of Congress Cataloging-in-Publication Data

Power, Max Singleton, 1942-
  America's nuclear wastelands : politics, accountability, and cleanup /
Max S. Power.
     p. cm.
  Includes bibliographical references and index.
  ISBN 978-0-87422-295-1 (alk. paper)
   1. Radioactive waste sites—United States. 2. Radioactive pollution—United States.
3. Radioactive waste disposal—United States. 4. Nuclear weapons—Environmental
aspects—United States. I. Title.
  TD898.15.P69 2008
  363.72'890973—dc22

Fine Quality Books from the Pacific Northwest

Photos in *America's Nuclear Wastelands* were provided courtesy of the U.S. Department of Energy.

Cover image: On July 6, 1962, Sedan crater at the Nevada Test Site was formed by a 100KT nuclear device (buried 635 feet underground) to test the applicability of atomic explosions in deepening harbors and building canals. Twelve million tons of earth were displaced, forming the 1,280-foot diameter and 320-foot deep crater.

For my sons, Stephen and Erik.

Contents

Acknowledgments...xi

Introduction... xiii

1. The Legacy of Nuclear Weapons Production............................1

2. Risk: Perception, Assessment, and Conflict...............................17

3. The Legal and Regulatory Basis for Cleanup 33

4 Politics, Jobs, and Public Engagement...49

5. How Clean Is Clean Enough?...67

6. The Geographic Chess Game..91

7. Cleanup Accomplishments 117

8. Cleanup Challenges for Today and Tomorrow 137

9. Trust and Momentum to Get the Job Done............................157

Index... 185

About the Author..193

Acknowledgments

So many people have contributed to *America's Nuclear Wastelands* that it is hard to know where to begin. About a decade ago, Dan Silver told me that I needed to write about "the end of the nuclear fuel cycle" before I "shuffled off this mortal coil." I am not sure this is the book Dan had in mind, but his remark stuck with me and goaded me along.

Many of my colleagues and fellow travelers on the journey described herein have contributed valuable insights, comments, and corrections to hazy memories. I am especially indebted to information resource specialist Valarie Peery, as well as John Price, Suzanne Dahl, and Melinda Brown, all of the Washington Department of Ecology's Nuclear Waste Program, and to Ken Niles and Dirk Dunning of the Oregon Department of Energy's Nuclear Safety Program. Terri Traub and Janice Parthree at the U.S. Department of Energy's public reading room in Richland spent many hours tracking down arcane and elusive documents and ephemeral tidbits of history.

Todd Martin, former chair of the Hanford Advisory Board and consultant at other nuclear sites around the nation, has been a constant sounding board and fount of knowledge. Kathleen Trever of the Idaho National Laboratory Oversight Office interrupted an extremely busy life in the midst of an unexpected Idaho gubernatorial transition to provide valuable insights and resources.

Dr. John E. Till, Dr. Jon Brock, Mary Lou Blazek-Smith, and Peter Beaulieu, a colleague from pre-nuclear waste days, all did their best to keep me intellectually honest. Fran Weinberg provided editorial acumen and invaluable advice about seeking publication.

The remarkably variegated members of Tuesday, a writers' group in Lincoln County, Oregon, served as guinea pigs to see whether this tale is told in a way that reaches a lay audience devoid of nuclear junkies. To the extent I avoid jargon, acronymitis, academic cant, and irrelevant technical discourse, the Tuesday writers largely deserve the credit. Moreover, their unflagging encouragement and support were essential to the completion of this book.

I am greatly indebted, as are many others, to two authors whose works are referenced herein and are highly recommended. No one can really comprehend Hanford's past and future without looking to Michele S. Gerber's *On the Home Front: The Cold War Legacy of the Hanford Nuclear Site* (1992, 2002), and Roy E. Gephart's *Hanford: A Conversation about Nuclear Waste and Cleanup* (2003).

Glen Lindeman, editor-in-chief at Washington State University Press, has been unflagging in his enthusiasm and support for publication of *America's Nuclear Wastelands.* He and the WSU Press staff have been very resourceful in providing background information, illustrations, and graphics that make the book more engaging, accessible, and useful to any reader.

Finally, my wife Marjorie has been my keenest, most hard-nosed editor throughout. Bless her for repeatedly correcting my "I'm trying to write a book about cleanup of the nuclear weapons complex," with "You ARE writing a book."

In the end, of course, I am solely responsible for the information and interpretations contained herein. If anyone takes issue, I hope she or he will do so in a way that stimulates discussion about the very challenging tasks we, as citizens, face in cleaning up the country's complex of nuclear weapons production facilities.

Introduction

I will show you fear in a handful of dust.
—T.S. Eliot, *The Wasteland*

In recent years, Americans have been treated to a number of excellent books describing atomic weapons development, the Cold War nuclear arms race, and the fate of J. Robert Oppenheimer and other key participants in those events. In contrast, my story is at the "other end" of America's wartime and Cold War era rush to produce nuclear weapons. *America's Nuclear Wastelands* focuses on contaminated soil and water, and the many risky and aging nuclear facilities scattered across the United States. The costs of cleaning up this legacy will be hundreds of billions of dollars.

I worked on nuclear waste and cleanup issues on behalf of the State of Washington for two decades. During that time, many well-educated, intelligent, serious people said to me: "I'm glad you're working on this. It's much too scary and complicated for me." This book is my response.

Of course these issues are scary and complex! We must decide how to retrieve and/or isolate radioactive and toxic materials that can harm people today and for many generations to come. The nuclear projects that left a legacy of contamination were rushed, secret, and pioneering. Their instigators could not have anticipated how these materials would travel in the environment, how the chemistry would change over time, and what eventually would be learned about health effects on humans.

Very few scientists and specialists actually "created" the processes to make the materials used in atomic and hydrogen bombs; a vastly greater number of people who constructed and operated nuclear weapons facilities were not given the overall picture and they were sworn to secrecy about the work they did. Consequently, the wartime and crisis atmosphere that allowed pell-mell development of nuclear weapons left a costly legacy of waste and contamination.

Today, many people have become involved in decisions about cleaning up this legacy—and still more need to do so. Indeed, cleanup actions will not be acceptable, sustainable, and protective in the future without the participation of people with a broad range of expertise, values, concerns, and insights. Only such broad participation based on openness, accountability, and trust can assure better outcomes as the nation continues to invest in the cleanup effort. Such engagement is important for three reasons:

- The technical challenges of cleanup require expert knowledge and special training, however, *the definitions of the overall problems and the decisions about how to proceed are matters of "public policy."* When policy choices reflect a broad range of values and perspectives, a better prioritization of issues and better decisions result. This means involving citizens who have different concerns, fears, and backgrounds. Moreover, the experience of recent decades indicates that lay citizens can master the necessary information and participate in achieving acceptable and sustainable results.

- *The ongoing nuclear-site cleanup effort provides an opportunity for Americans to similarly develop the knowledge and skills to deal with other complex, technologically-based, long-running public policy challenges.* If nothing else, this is a cautionary tale about what happens when, as in the case of the Manhattan Project and the Cold War, crash programs are launched in relative isolation and led by a narrowly-technical group of experts, who can give little attention to potential long-term negative consequences. Faced with global warming, a decreasing availability of fossil fuels, and other urgent problems, our nation today might easily acquiesce in similar approaches—with equally undesirable long-term results.

- *The governmental institutions and personnel charged with meeting cleanup challenges, when left to themselves, do not do a good job of engaging a broad range of citizens.* The dominance of the federal government in this arena tends to weaken the customary checks and balances of American federalism. For a variety of historical, legal, and political reasons, the U.S. Department of Energy—today's responsible agency—is inclined to a "we're the experts, we know best" attitude as it pursues cleanup. Accountability to a concerned public will not happen unless people actively pursue it.

In the 1970s and 1980s, the concerned public—led by committed and curious people, including both the technically-trained and the non-technical—diagnosed problems resulting from insulated and secretive nuclear weapons production complexes located across the country. They were able to elevate these problems on the public agenda, and push for better, more thoughtful, and more complete solutions than otherwise would have occurred. The battle is not over, as the reader will see, and continued involvement by ordinary lay citizens is crucial to continued success.

In *America's Nuclear Wastelands*, I try to provide the concerned citizen with encouragement, presenting guidelines for engagement in non-technical language and by presenting pertinent examples. I attempt, as well, to describe the institutional and political environment in which this engagement takes place. The political battles over cleanup have many dimensions. Perhaps the most obvious is the clash between those who trust the experts to get on with the job, and those who for a variety of reasons insist on a broader public process. The struggles over nuclear cleanup illustrate this fundamental divide in American political culture

As a political scientist who came of age in the mid-20th century, I have been fascinated by the political dynamics surrounding the cleanup of nuclear weapons sites. There is a complex interplay between technical perspectives and geographically-based political action. Initially, I was surprised to find conservatives—individuals generally committed to limited government and supporting states' rights—on the side of unchecked federal power and against state involvement in this arena. Traditional liberal Democrats, on the other hand, often appeared as advocates for states' rights and limits to federal power.

In the mid-1980s, for example, Democratic members of the Washington House of Representatives from districts west of the Cascade crest passed legislation requiring the state to issue permits to carriers transporting radioactive waste to and from Hanford. The bill died quietly in the Senate Transportation Committee, chaired by a Republican whose eastern Washington district included a substantial portion of the Hanford nuclear reservation.

The real differences, however, were—and are—not primarily partisan. The public's perception of risk differs from that of experts in the nuclear field. And perceptions of risk relate to jobs—people in communities that have grown up around and depend on nuclear facilities view risk differently from the broader public. This fundamental divide plays itself out between leaders representing constituencies where opposing perceptions prevail.

Perceptions, indeed, generate a great deal of conflict and consternation. In the late 1980s, I testified at an Idaho joint legislative committee hearing concerning the formation of the Idaho National Laboratory Oversight Office. I was asked to talk about Washington's experience when we developed our regulatory role in regard to Hanford. The senator chairing the hearing made it clear to me that the large dimension of Washington's program was of some concern. He hoped to make the case that a small number of additional positions in Idaho's regulatory agencies would be adequate to do the job in his state. During my testimony, a member specifically asked about the relatively large size of Washington's program. I said, among other things, that the sheer scope of contaminated sites and the larger number of different waste problems at Hanford would suggest that Washington's program be more heavily staffed than Idaho's.

Moments after the hearing, an activist leader of the Snake River Alliance accosted me in Boise's capitol rotunda, berating me for daring to suggest Hanford's problems were worse than those at the Idaho National Laboratory. The next morning, Hanford's deputy manager called me, greatly vexed over the same testimony. It seems his counterpart at INL called him right after the hearing to gloat because I had said the INL had done a better job managing its wastes than Hanford.

Obviously, this is not a policy realm for the conflict averse. However, I have written this account with the same conviction as one of Thomas Jefferson's precepts. In an 1815 letter to a colleague, Jefferson stated: "Difference of opinion leads to enquiry, and enquiry to truth; and I am sure... [we] value too much the freedom of opinion sanctioned by our Constitution, not to cherish its exercise even where in opposition to ourselves."

People Matter

In my view, involvement with cleanup issues provides an opportunity to show that Americans can work together civilly and productively to solve complex problems. *America's Nuclear Wastelands* is based on my own journey through the world of nuclear waste issues and incorporates my personal experiences, observations, and reflections.

At bottom, this is a story about people—people in their daily jobs, people pursuing their passions, people seeking understanding or redress. Hundreds and hundreds of individuals have figured in my journey. I have chosen to sketch the story in broad terms to provide a guide to others who will, I hope, choose to become involved. This broad scope, regretfully, means that I am not able to talk about or credit many of these individuals in the space allowed here.

Five people who departed this life as I made that journey, however, especially displayed the traits essential for successful public engagement. They represent the range of types of participants in the policy debate over nuclear waste. Varied as they were,

- All respected and valued the opinions and contributions of people with markedly different views from their own.
- All made it clear that they expected openness and accountability from themselves and others.
- All persisted in working with people with different perspectives and goals to find common ground to make the world safer for us all.

In alphabetical order:

Dr. Richard Belsey. A retired Portland physician and teacher, Dick Belsey was a passionate member of the Physicians for Social Responsibility. A charter member of the Oregon Hanford Waste Board, the Hanford Future Site Uses Working Group, and the Hanford Advisory Board, he pioneered public engagement in cleanup issues. Dick had strong views about the dangers of plutonium and nuclear proliferation, and the potential clinical effects of radiation on real people. Yet, as chair of a Hanford Advisory Board committee, he brought DOE and contractor personnel together along with people of all other persuasions to work out practical and acceptable approaches to overcome the complex and nagging legacy of Cold War plutonium production. Dr. Belsey also illustrated for me the importance of "kicking the tires"—of periodically looking at the detailed implementation of a broad policy to assess its effectiveness.

Lori Friel. As staff attorney at the Western Interstate Energy Board (WIEB), Lori Friel led work by the board's High-Level Radioactive Waste Committee in dealing with the transportation of nuclear waste in the western United States. The committee included policy people from the various states (such as myself when I served on the committee) and emergency management and law enforcement officials. Included were members from such states as Nevada and Washington, targeted for waste import, and states like California and Colorado, eager to export waste. The committee's activities are funded by cooperative agreements with the Department of Energy, which expects that the work be of value to DOE. Lori consistently brought together individuals with conflicting viewpoints to propose reasonable measures making nuclear waste transportation safe, and for the concerned public to view it as being safe, too. Strengths that Lori brought to the task were her ability to provide clear legal and technical analyses and syntheses, and her insistence on accountability. Lori wanted to fully understand—and to help others understand—why decisions were made. Her work led to, among other things, a widely hailed transportation safety program for hazardous material destined for the Waste Isolation Pilot Plant in New Mexico.

Terry Husseman. As director of Washington's Office of Nuclear Waste Management, and later assistant and deputy director of the state's Department of Ecology, Terry Husseman was an advocate for mediation and reconciliation. In the earlier position, he was my boss. Terry had a marked, if not always obvious, influence on nuclear waste issues and the Hanford cleanup. Terry practiced and promoted three important principles:

- More public involvement in decision-making leads to better results.

- Do not focus on the negative—telling people what they have done wrong. Point them in the direction of doing the right thing.
- Trust someone unless he or she clearly violates your trust.

Herman Reuben. A jovial though quiet man, Herman Reuben was a member of the Nez Perce Tribe and served for many years on the tribe's environmental staff, funded by various Department of Energy related projects. Herman developed newsletters and displays that made tangible the connection of tribal people to the land. He patiently exhibited and explained how roots, seeds, and berries were traditional sources of food, medicine, and meaning. Herman worked amiably with representatives of other tribes, some of whom had historic long-term issues with his own tribe, and with narrowly focused white technicians, as well as white doctors and bureaucrats. He strived to help Indian people understand how Hanford's activities may have affected—and could still affect—their health.

Jon Yerxa. In the early 1990s, Jon Yerxa was the portal between DOE's Richland Operations Office and an increasingly involved stakeholder community. This was not an easy time, as the Department of Energy was increasingly on the defensive when shifting an ingrained culture away from weapons production toward cleanup. Of these five people, Jon undoubtedly was the most frustrating and maddening from my point of view. And yet, he was extremely focused on two key aspects of his role. He had to tell his hunkering-down management how much they really needed good public involvement. He also had to reveal to the stakeholder community, with as much candor as possible, the difficulties and constraints faced by his management in moving toward openness and public involvement. Jon always was ready to try *something*, when someone else in his difficult position could have easily retreated into bureaucratic obfuscation.

* * *

Many other people, regretfully mostly unnamed in the account that follows, also displayed the fine qualities of these individuals. I hope that these glimpses will help the reader sense the importance of individuals in the big picture of nuclear waste cleanup.

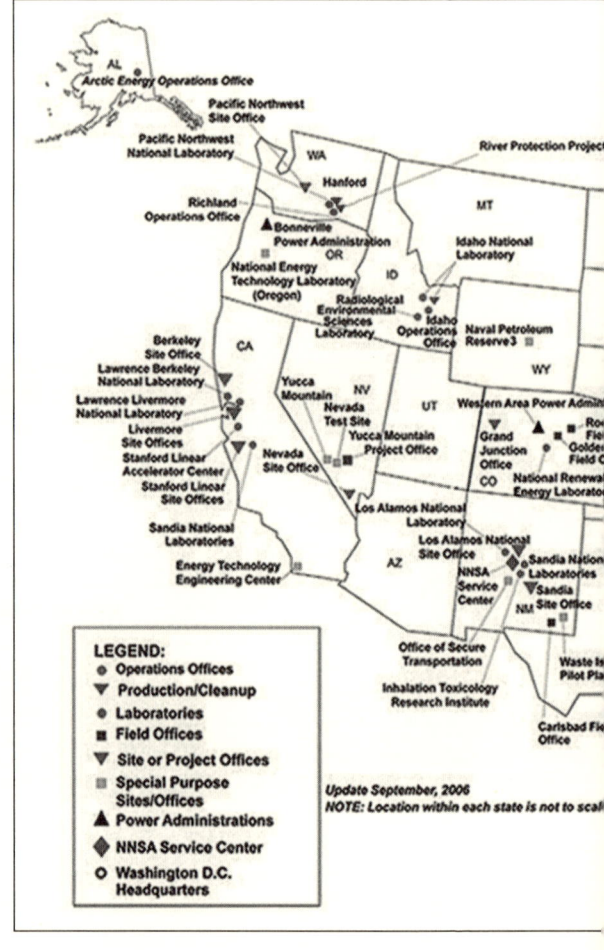

Hanford's B-Reactor in operation in the 1940s.

1

The Legacy of Nuclear Weapons Production

Birth of the Atomic Weapons Complex

I f you stand atop Gable Mountain, on the Hanford Site in south-central Washington State, you can see forever. The Columbia River forms a broad "h"-shaped bend and you are standing under its crook overlooking a sage-brush covered plain. Also visible to the north, Wahluke Slope and the Saddle Mountains form the horizon; to the west and southwest stand the Rattlesnake ridgeline and hills—claimed by area old-timers to include the highest mountain in the "lower 48" without trees. In fact, trees are very scarce in the sweeping vista. On a clear day, snow-covered Mount Adams (12,276 ft.) in the Cascade Range is visible, over a hundred miles to the west.

Had you stood here in early 1943, you would have seen the small agricultural settlement of White Bluffs to the north, a few farms to the east and southeast, with the hamlet of Richland far beyond to the south. By early 1945, however, this landscape had been dramatically transformed by the "Hanford Engineer Works." From this vantage point, you would have seen new major industrial facilities along the river to the north and west, and still more quite close by your feet, to the south and southwest. Hundreds of structures, tent and trailer camps for soldiers and construction workers, and roads, rail lines, and electrical transmission lines would sprawl in all directions. Several vent stacks over 200 feet high would redefine the horizon.

You might well have observed a similar transformation had you been looking over East Poplar Creek from Chestnut Ridge in east

Tennessee in those same years. The Poplar Creek and Bear Creek watersheds suddenly sprouted massive industrial complexes—in this case, seeming to overwhelm the "creeks and hollers" character of the original landscape.

More than 130,000 people would be at work constructing and operating these and other nuclear related complexes across the nation. These were the first industrial-scale facilities erected to capture atomic energy—in this case for destructive purposes. Deep into World War II, and faced with the threat of new and terrifying weapons developed by Nazi Germany, the United States government embarked on the Manhattan Project[1] to develop the atomic bomb. The build-up was extremely rapid, the process riddled with uncertainties, and the whole effort known to, overseen, and guided by only an extremely small group of scientists, government officials, and military officers. The strictest secrecy was the norm.

Under wartime pressure, the decision was made to pursue two separate approaches to making atomic bombs. One relied on the fission of enriched uranium—a process requiring the construction of massive magnetic-accelerator and gaseous-diffusion facilities at Oak Ridge, Tennessee, to produce the uranium that obliterated Hiroshima. The second process relied on plutonium as the fissile material. Plutonium would be made by irradiating uranium in nuclear reactors (or "piles"), then separating the material chemically. Fuel-fabrication, atomic piles, and separation facilities sprouted up along and near the Columbia River's Hanford Reach, producing the plutonium that fueled the Trinity (New Mexico) and Nagasaki bombs.

These were only two of hundreds of places around the country that would be transformed by the commitment to nuclear weapons, and subsequently to producing public electricity and providing nuclear power to propel ships.

I will leave the story of the adverse impacts of the mining and milling of uranium to others—people in many mining communities in the arid West can tell you how significantly this aspect of the atomic era affected them. Once refined uranium was available, it found its way to many of the new nuclear facilities. In the early days, Hanford and Oak Ridge were the largest, but by no means the only ones.[2]

Hanford's single-shell Tank C-106 under construction in 1943. Six decades later, it was the first of Hanford's 177 underground storage tanks selected for major fieldwork preparations to retrieve hazardous waste and for closure.

When the Soviet Union became a nuclear power in 1949, ushering in the Cold War, dormant weapons-making operations at these two big sites were revived and new facilities added. Hanford increased from three reactors in 1946, to eight by 1955, dispersed in wide intervals along the south side of the Columbia. A ninth, the N-Reactor, was added in 1963, a result of President John F. Kennedy's pledge to "close the missile gap" in the nuclear arms race with the Soviet Union.

By the early 1950s, naval nuclear propulsion spawned major new facilities on a gunnery range in southeastern Idaho at the Idaho National Laboratory.[3] Uranium enrichment operations spread from Oak Ridge to Portsmouth, Ohio, and Paducah, Kentucky. Additional uranium-related activities sprang up at such places as Fernald, Ohio, near Dayton, and Weldon Spring, Missouri, just outside St. Louis. Hanford's plutonium production capabilities were duplicated at Savannah River in South Carolina. The machining of plutonium for nuclear weapons became a special enterprise at Rocky Flats, between Denver and Boulder, Colorado. Along the way, 1,350 square miles of southern Nevada's desert emerged as the nation's principal nuclear weapons test site.

Several factors drove the locating of these operations. For the earliest facilities, such as Hanford and Oak Ridge, as well as Los Alamos in New Mexico, remoteness and the ability to secure the sites from intrusions were critical. For the industrial facilities at Oak Ridge and Hanford, the availability of large amounts of reactor-cooling water and electrical power also were key factors. Water and electricity were provided by both the Tennessee Valley Authority and the Columbia River, respectively. As the Cold War took shape, the government also pressed to have redundant, geographically-dispersed facilities. If one site were obliterated by a Soviet missile or bomber nuclear strike, the capability to produce weapons remained elsewhere.

Politics increasingly came to play a part as well; nuclear facilities brought high-paying construction jobs, followed by higher-paying technical employment. It is said that the massive investment at Oak Ridge—the "Clinton Engineer Works"—during World War II was the price President Franklin D. Roosevelt paid for the support of Senator Kenneth McKellar of Tennessee in diverting huge appropriations to the Manhattan Project. By today's environmental standards, the Oak Ridge facilities were poorly sited. When I stood on Chestnut Ridge more than a half-century later, I could see hundreds of workers' cars in the parking lots and understood how politically difficult it would be to shut down the Y-12 plant, even though the country was deemed to have a surplus of highly enriched uranium.

Atoms for Propulsion, Deterrence, and Peace

After World War II, the Atomic Energy Commission (AEC) took over the complex of facilities developed by the Manhattan Project. In the U.S. Congress, the Joint Committee on Atomic Energy had exclusive, largely secret, oversight of the AEC's budget and activities. The AEC's mandate was broader than weapons production—it was to promote other applications of nuclear power, both for defense and in the civilian sectors; the latter were formalized under President Dwight D. Eisenhower's designation, "Atoms for Peace."

These activities added new missions and waste-producing activities at existing sites, thus complicating the eventual cleanup and disposal problems. Moreover, they were carried out within a cultural framework—born during the Manhattan Project—characterized by extreme secrecy, compartmentalization, relatively little fiscal accountability, and scientific and technical isolation (some call it arrogance). There was continuing pressure to produce results, without much regard for costs and consequences.

Relics of this era, both physical and attitudinal, continue to lurk about. One can visit a rusting experimental nuclear aircraft engine parked on southeast Idaho's arid plain. A tower for launching a nuclear rocket still looms over the western plateau of the Nevada Test Site. A few miles to the east, one can peer into the Sedan crater. This massive hole was created to see if nuclear explosions could be used to deepen harbors and build canals. In 1961, "Project Gnome" detonated a nuclear device in an old New Mexico salt mine, attempting to store and use the heat for generating electric power—ironically laying the groundwork for today's successful Waste Isolation Pilot Plant, east of Carlsbad, New Mexico.[4]

A few examples from my own experience illustrate the nature of the closed culture that persisted well past the end of the Cold War. In 1986, the U.S. Department of Energy (the successor agency to the Atomic Energy Commission in handling operational and developmental activities) took a significant step toward openness when producing a draft environmental impact statement for the management of defense wastes at the Hanford Site. The draft, however, contained no analysis of the impacts of the hazardous "chemical" components of Hanford's defense wastes, as opposed to the "radiological" components. This was a decade after Congress passed the Resource Conservation and Recovery Act—the legal basis for defining and requiring proper management of hazardous chemicals in the United States.

In the mid-1980s, state and local officials on a bus tour at Hanford encountered a low-boy truck on a highway hauling a heavy round container. The truck was preceded and followed by Hanford Patrol vehicles. When asked what this shipment might be, the guide told the group: "It's nothing. You didn't see this."

In the early 1990s, the independent Technical Steering Panel for the Hanford Environmental Dose Reconstruction Project[5] secured a commitment from Secretary of Energy James Watkins that information about radiation released into the environment prior to 1970 would be declassified and made available to the panel and the public. The multi-disciplinary panel included (nuclear) health physicists, meteorologists, environmental scientists, statisticians, cultural anthropologists, public participants, and Native American tribal members.

The custodians of the information at Hanford's Pacific Northwest National Laboratory—those responsible for the complex declassification process—nonetheless resisted the concept that the panel members, let alone the public, had a "need to know" this information. They further insisted that those who generated the information had the right to determine whether those requesting it had a legitimate "need to know."[6]

In 1962, reactor-effluent discharges saturated the groundwater of Hanford's 100-K Area with hexavalent chromium.

The Legacy of Waste and Contamination

At the end of the Cold War, the federal government began to inventory the wastes and contamination left by 45 years of weapons production and nuclear operations and research. The Department of Energy (DOE or USDOE), as the legatee of the problem, established the Office of Environmental Management in 1989. Since then, Environmental Management's annual budget has fluctuated around $6 billion. In order to help Congress and the public grasp the extent of the cleanup tasks, the office has published reports and environmental statements, often at the specific direction of Congressional committees. The following brief sketch of the nature and extent of the nation's waste and contamination problem is based on this body of work. After more than a decade of assessment, the Environmental Management program estimated that it would require as much as $212 billion and 70 years to clean up the legacy of nuclear waste and contamination at 113 sites.[7] The office summarized the extent of the problem:

- 1.7 trillion gallons of contaminated groundwater, an amount equal to approximately four times the daily U.S. water consumption.
- 40 million cubic meters of contaminated soil and debris, enough to fill approximately 17 professional sports stadiums.
- The storage and guarding of more than 18 metric tons of weapons-usable plutonium, enough for making thousands of nuclear weapons.
- Managing over 2,000 tons of intensely radioactive spent nuclear fuel, some of which was corroding.
- Storing, treating, and disposing of radioactive and hazardous waste, including over 160,000 cubic meters currently in storage, and over 100 million gallons of liquid, high level, radioactive waste.
- Deactivating and/or decommissioning about 4,000 facilities no longer needed to support active DOE missions.
- Implementing critical nuclear non-proliferation programs for accepting and safely managing spent nuclear fuel—containing weapons-usable highly enriched uranium—from foreign research reactors.

- Long-term care and monitoring—or stewardship—for potentially hundreds of years at an estimated 109 sites following cleanup.

A watchdog group, Resources for the Future, in 2000 summarized the problem this way:

> The nuclear weapons complex comprises 3,750 square miles and, although the overwhelming majority of this land is uncontaminated (less than 15% of the land at the five major sites is contaminated), the contamination that does exist presents difficult technical challenges because of the presence of radionuclides. Seventy-five million cubic meters of soil are contaminated, enough to cover the entire island of Manhattan more than five feet deep. There is currently no effective technical solution for remediating much of the 1.8 billion cubic meters of contaminated groundwater in the complex (enough to cover Manhattan with 135 feet of water).[8]

In early 2002, the George W. Bush administration estimated that it could shave 35 years and $50 billion off the estimated cost and time for cleanup. The reductions could be achieved by leaving more waste and contamination in place than the earlier estimates anticipated, and by pursuing a more aggressive contracting approach. However, neither this, nor the previous administration's estimates, could be regarded as precise.[9]

Providing an accurate and complete picture of the nuclear legacy was not easy in 2000—and still remains difficult today. The nation's nuclear program may have begun as a primarily military enterprise, but by the 1980s, it had evolved into anything but a centralized endeavor. The initial secrecy and compartmentalization that characterized the Manhattan Project, however, continued as the nation's complex of nuclear facilities and missions expanded; so, too, in the government's contracting approach. Large companies—DuPont initially—were hired under fairly minimal governmental supervision to "get the job done."

Therefore, a whole series of "operations offices" grew and controlled activities at one or more nuclear facilities and sites—in Nevada, Idaho, and Ohio; at Chicago, Albuquerque, Richland, Oakland, Oak Ridge, and Savannah River; and so on. Each field office hired one or more separate contractors to operate their sites, and each had its own rules and procedures, while contractors developed their own protocols to implement them. After all, the contractors, not government officials, actually were managing and operating the many technically-complex and often dangerous facilities. There was no regulation outside the oversight provided by the Atomic Energy Commission and its successor agencies. The national laboratories, particularly those most directly involved in weapons development, such as Lawrence Livermore in California and Los Alamos, became powers unto themselves.

As a result, problems, activities, and processes were defined differently, data were kept under various labels and for differing lengths of time, the level of detail found in reporting systems varied widely from site to site, and the responsibilities for assuring the reliability of information were diffuse. Federal operations offices—actually, often their contractors acting on their behalf—were the custodians of records, which were not sent to a central point.

From 1989 and even into the first years of the 21st century—after Congress finally had demanded new accountability and action regarding nuclear waste cleanup issues—DOE headquarters officials continually were engaged in efforts to extract information from the department's own many reluctant "field elements." What outwardly appeared to be a command and control system actually was widely characterized by a mixture of DOE's pleading and frustration from Washington, D.C., on the one hand, and resistance and evasion in the field offices on the other.[10]

What Does "Cleanup" Mean?

What are the activities that could require $212 billion and 70 years to accomplish? At its core, "cleanup" is a simple concept: Isolate contaminants so that they will not reach human beings through the environment *in concentrations large enough to harm people.* Most of the "cleanup" does so in one of three ways:

- Immobilize contaminants that are presently mobile or easy to mobilize—e.g., target liquids or materials that dissolve in water, vaporize in the air, or that might migrate through the soil or groundwater to plants that animals and humans consume.

- Move contaminants into disposal facilities that are relatively impenetrable to groundwater, plants, animals, and human intruders.
- Build barriers over and/or around contaminated areas to prevent intrusion or migration.

The good news, insofar as radioactive contaminants are concerned, is that radioactive elements do decay away. As a general rule, the more energetic and penetrating they are, the shorter their half-lives. Iodine-131, for example, has a half-life of about 8 days. In large doses, I-131 concentrates in the thyroid gland and may damage tissue. I-131 was the principal cause of widespread public (as opposed to worker) health impacts from the Chernobyl disaster in the Ukraine. From 1945 to 1951 in central Washington, too, local populations were

Chernobyl

In the Soviet Union on April 25, 1986, a sudden massive power surge in Unit 4 at the Chernobyl nuclear power plant produced two explosions. The first lifted the cover plate off the reactor, and both explosions released large amounts of radioactive material into the atmosphere. The reactor's design did not include a sealed containment outer structure.

The accident occurred when the facility's crew, attempting to test the reactor's productivity under abnormal conditions, disabled the automatic shut-down devices. The explosions and subsequent fire released all of the reactor's xenon gas (half-life, 5.2 days), half of the iodine-131 and cesium-137 (half-life, 8 days and 30 years respectively), and 5% of the remaining radioactive material. The lighter radioactive materials were carried in the atmosphere over parts of the Ukraine, Belarus, Russia, and Scandinavia. Residents within 30 kilometers (19 miles) of the plant were evacuated and resettled.

Some 30 first responders shortly succumbed and died from radiation exposure; another 19 died subsequently. Ongoing health studies have found heightened levels of leukemia among those who responded to the emergency and participated in the cleanup, as well as increases in thyroid cancer among children who were living in the area at the time.

See the Australian Uranium Association briefing paper, "Chernobyl Accident," May 2007, at www.uic.com.au/nip22.htm. This Web site also includes a number of useful references and links to international agencies.

exposed to iodine-131 and other radionuclides released directly into the environment by Hanford operations.

Cesium-137 and strontium-90, highly penetrating radioactive isotopes resulting from fission in nuclear reactors, have half-lives approaching 30 years. If one accepts the maxim that isotopes are no longer particularly dangerous after 10 half-lives, then the isolation of cesium and strontium from humans and the environment is necessary for about 300 years.

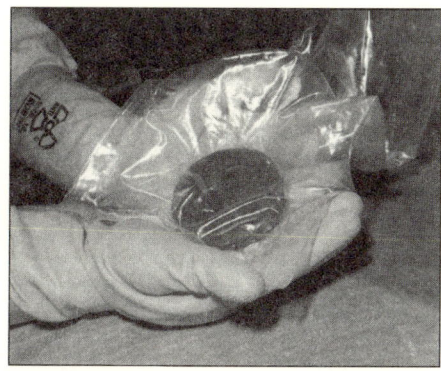

A worker handles a plutonium button, Mark 31, at the Savannah River facility, South Carolina, December 31, 1997 (Environmental Management program).

However, there are at least three aspects of bad news:

- Other isotopes have extremely long half-lives, and, while they may not be highly penetrating, they are very toxic if ingested or inhaled. Plutonium-239, used in weapons production, for example, is highly toxic if internalized, and has a half-life of 24,110 years (given the 10 half-lives maxim, hazardous for over 240,000 years). Pu-239 is one of the chief products of the Cold War arms race.
- Even if the radiological risk decays away, many wastes and by-products of nuclear activities contain chemical components that are toxic and do not decay. These include heavy metals such as lead, zinc, and cadmium, and other known hazards like mercury and beryllium.
- Many of the early practices in the nuclear age did not clearly recognize the need to isolate contaminants when they were put in disposal. At Hanford, for example, inadequacies in the disposal of billions of gallons of wastewater from chemical separation processes means that contaminated soil and groundwater now extend over more than 100 square miles. Below Oak Ridge, the injection of contaminants into fractured strata of rock saturated with groundwater—"reverse wells"—has become one of DOE's more intractable cleanup problems. Early in the nuclear era, highly contaminated wastes were dumped off the edge of the mesa where

Vitrified hazardous material in amber-colored glass from the Fernald Closure project, Ohio, August 10, 1995.

Los Alamos is situated and into Pueblo Canyon, contaminating the stream below—a tributary of the Rio Grande River.

The cleanup effort covers a wide range of problems and situations. However, most activities are directed toward the solidification and engineered disposal of harmful materials, and the containment of hazardous wastes already dispersed in soil or groundwater, thus keeping it from spreading further. For example:

- Highly-radioactive, hazardous, liquid waste and sludge in underground storage tanks at Hanford, Savannah River, and New York's West Valley have been, or will be, vitrified—i.e., turned into glass and encased in steel cylinders. Less radioactive wastes extracted from these tanks, as well as other liquid or loose dry wastes, are being grouted in cement inside steel containers.
- Highly radioactive vitrified wastes, together with spent nuclear fuel[11] rods, are, as a matter of law and policy, to be disposed of in a deep geologic repository located in a stratum of welded volcanic rock (called tuff), several hundred feet beneath Yucca Mountain, Nevada. Plutonium-contaminated wastes, on the other hand,

are sent to the Waste Isolation Pilot Plant, a series of caverns in a salt bed some 2,000 feet under an arid plain east of Carlsbad, New Mexico. Other radioactive wastes that are relatively less long-lived and/or toxic go to a few other designated disposal sites, or are buried in engineered disposal cells at the sites where the wastes were generated.

- In situations where historic practices have caused the dispersion of contaminants into the soil and groundwater, a typical containment strategy includes "pump and treat"—i.e., pumping contaminated groundwater to the surface and removing as much of the contamination as possible; inserting underground barriers to prevent groundwater from moving in or out of contaminated zones; and "capping"—constructing surface barriers to prevent infiltration by water, plant roots, or burrowing animals.

Nuclear Fear and Federalism: Cleanup Impetus and Issues

Even before the Soviet Union collapsed in 1991, two factors strongly impelled the federal government to address the legacy of nuclear waste and contamination. First, American public opinion regarding nuclear power and weapons took a decidedly negative turn after the Three Mile Island plant failure in 1979, and

Three Mile Island

Early in the morning, March 28, 1979, a series of malfunctions led to a loss of cooling water in the Unit 2 reactor vessel at Three Mile Island, a commercial nuclear power plant near Harrisburg, Pennsylvania. As a result of overheating, about half of the nuclear fuel in the reactor core melted.

Over the next few days, officials were extremely concerned that the thermally and radioactively hot material would rupture Unit 2's vessel and perhaps breach the reactor's containment building. Had this occurred, a large release of radioactivity into the atmosphere would have affected thousands, or even millions, of people. In fact, most of the radiation released in the accident was contained.

The Three Mile Island incident is regarded as the most serious nuclear reactor accident to ever occur in the United States. Nonetheless, nearly all follow-up studies have found very low radiation doses, and no discernable health effects as a result of the accident.

See, U.S. Nuclear Regulatory Commission, "Fact Sheet on the Three Mile Island Accident," at www.nrc.gov/reading-rm/doc-collections/fact-sheets/3mile-isle.html.

became still more negative as a result of the Chernobyl disaster in 1986.[12] Second, based on court decisions and Congressional amendments to environmental laws in the mid-1980s, the states began to exert authority in regulating Department of Energy facilities and practices.

As the cleanup effort gained momentum, specific decisions and actions to carry out cleanup activities have raised several difficult public policy issues. These include:

- *How clean is clean?* What is a safe (or reasonable) prevailing level of risk, and who has the authority to say so?
- *Should contamination be removed or left in place?* How should we balance short-term risks to workers and costs to the taxpayer for early removal and treatment, against the long-term risks and costs of leaving contaminants in situ?
- *If wastes are to be removed, then to (and through) whose backyard should it go?* How does our federal system function in the face of this part of the nuclear legacy? What are some key characteristics of the politics of dealing with nuclear waste transportation and storage?
- *How do the affected communities around nuclear facilities balance jobs, risks, and future activities?*

In the next two chapters, I will provide some background about how risk is measured and perceived, followed by a brief sketch of the legal and regulatory environment in which cleanup takes place. With this background in place, I then will turn to the political and institutional dynamics that drive how these issues are addressed. Throughout, I hope to demonstrate the importance of having the public play a major part in resolving these issues.

Notes

1. The federal appropriation for atomic weapons development was hidden in a budget for the U.S. Army Corps of Engineers' Manhattan Engineering District—thus the derivative for the title, "Manhattan Project." The most widely read history regarding the Manhattan Project is by Richard Rhodes, *The Making of the Atomic Bomb* (New York: Simon and Schuster, 1986).

2. For a comprehensive view of the various sites and processes involved in nuclear weapons production, see USDOE, Office of Environmental Management, *Linking Legacies: Connecting the Cold War Nuclear Weapons Production Processes to Their Environmental Consequences* (Washington, D.C., January 1997).

3. Established in 1949, the Idaho National Laboratory (INL) has undergone several name changes over the years. In 1977, it was re-designated as the Idaho National Engineering Laboratory (INEL); in 1997, the Idaho National Engineering and Environmental Laboratory (INEEL); and in 2005, resumed its original name, the Idaho National Laboratory (INL). In *America's Nuclear Wastelands*, I usually apply the latter designation.

4. Chuck McCutcheon, *Nuclear Reactions: The Politics of Opening a Radioactive Waste Disposal Site* (Albuquerque: University of New Mexico Press, 2002), pp. 23–24. I will more fully discuss the Waste Isolation Pilot Plant (WIPP) in later chapters.

5. The Hanford Environmental Dose Reconstruction Project was established to estimate off-site radiation doses received by the public during the earlier years of Hanford's operations, in order to provide a basis for determining whether there had been adverse health impacts. See Ken Niles, *Reconstructing Hanford's Past Releases of Radioactive Materials: The History of the Technical Steering Panel, 1988–1995* (Salem: Oregon Office of Energy, November 1996).

6. To be fair, once the initial resistance was overcome, the Richland Operations Office and the Pacific Northwest National Laboratory performed admirably in declassifying and making publicly available a mountain of historical records. By the end of the 1990s, Hanford's declassification program won wide praise from both concerned citizens and groups.

7. USDOE, Office of Environmental Management, *Status Report on Paths to Closure* (Washington, D.C.: March 2000), pp. 1–2. In 1998, a Brookings Institution study estimated that the cost for atomic weapons from 1940 to 1996 had been $5.5 trillion in constant 1996 dollars; *Atomic Audit: The Costs and Consequences of U.S. Nuclear Weapons since 1940* (Washington, D.C.: Brookings Institution, 1998). The estimated cleanup price, therefore, would be under 4% of the overall costs of the weapons program.

8. Katherine N. Probst and Adam I. Lowe, *Cleaning Up the Nuclear Weapons Complex: Does Anybody Care?* (Washington, D.C.: Resources for the Future, January 2000), pp. 1–2.

9. U.S. Government Accountability Office (GAO), *Nuclear Waste: Better Performance Reporting Needed to Assess DOE's Ability to Assess the Goals of the Accelerated Cleanup Program* (Washington, D.C., July 2005), pp. 2–3. The GAO, however, concluded that the savings would be difficult to achieve in light of recent experience. Based on its initial assessment, the Bush administration reduced the cost estimate for the Environmental Management program from about $190 billion to about $145 billion in its fiscal 2004 budget request to Congress. Four years later, however, the estimate had risen to about $160 billion. See USDOE, *FY 2007 Budget Request: Environmental Management* (February 2006), pp. 48–49.

10. The Government Accountability Office (formerly the General Accounting Office) examined the confusing lines of authority and accountability at the Department of Energy in many reports, including the one cited above in 2005. Perhaps the

most trenchant analysis of the problem was produced by the President's Foreign Intelligence Advisory Board in a report on security problems at the Los Alamos National Laboratories: *Science at Its Best; Security at Its Worst* (Washington, D.C., June 1999). As early as 1989, the Natural Resources Defense Council and a host of environmental organizations sued the Department of Energy in order to acquire a consistent and complete inventory of wastes and contamination; *Natural Resources Defense Council v. Watkins*, Civ. No. 89-1835.

11. "Spent" nuclear fuel is defined as some form of uranium fuel that has been used in a nuclear reaction. At some point, depending on the purpose of the reaction (e.g., to produce heat to generate electricity, or to create plutonium for weapons), fuel becomes no longer particularly efficient. However, its use in nuclear reactions has produced many fission products, such as cesium and strontium, that are highly radioactive. Spent fuel also contains isotopes of uranium, plutonium, and other potentially dangerous materials.

12. Eugene A. Rosa and William R. Freudenburg, "The Historical Development of Public Reactions to Nuclear Power: Implications for Nuclear Waste Policy," in Riley E. Dunlap, Michael E. Kraft, and Eugene A. Rosa, eds., *Public Reactions to Nuclear Waste: Citizens' Views of Repository Siting* (Durham, N.C.: Duke University Press, 1993), pp. 33–63.

2

Risk: Perception, Assessment, and Conflict

The 24 Command Fire: An Example

About 3:30 p.m. on June 27, 2000, I was met by a hot wind when stepping outside of Washington State University's library near the Columbia River in Richland. Some 30 miles to the northwest, a high column of rising smoke in an otherwise clear sky made me change my planned route back home to Olympia, the state capital. The 24 Command Fire had just begun.

About two hours before, a driver had been returning to Richland from a dental appointment in Yakima when her car swerved across the Highway 24 centerline, colliding with a truck carrying apple waste. The truck's fuel tank burst and set grass afire next to the road. Within 72 hours, the blaze swept across some 160,000 acres of sage and grass lands, mostly on uncontaminated portions of the Hanford Site.

Within six weeks, laboratory analyses of air samples taken at the time of the fire showed elevated levels of plutonium and cesium, both on the Hanford Site and in nearby communities. The plutonium and cesium (not naturally-occurring elements) were likely residuals from the 45-year period when Hanford created two-thirds of the plutonium for America's nuclear weapons.

The fire had not burned any of Hanford's nuclear production facilities. Hard-working firefighters and existing fire-breaks, such as roads, rail lines, and graveled areas surrounding buildings, had prevented—but only just—the spread of much more contamination.

Two weeks after the fire, I visited burial ground 618-10, about 15 miles north of the WSU Tri-Cities campus from where I had seen the

Corroded spent nuclear fuel being moved underwater at the Hanford Site, May 7, 1996.

fire's beginning. This burial ground contained extremely radioactive wastes. The vegetation around it—but miraculously not on top of it—had been burned. The blaze also had swept across the dry plains up to a railroad and highway bordering on the 300-Area, where tinder-dry 1940s and 1950s buildings containing a wide range of chemical and radioactive contaminants stood on nearly 1,000 acres. The fire also had spread up to the edge of one of Hanford's infamous underground high level waste storage tank farms, located some 20 miles to the northwest.

DOE, EPA, and state health department officials assured the public that the elevated levels of plutonium and cesium did not come close to exceeding health-based standards that indicated people faced increased risk of cancer. But officials also admitted that they did not know the source of the contamination mobilized by the fire.

The *Seattle Post-Intelligencer* later reminded readers that, at the time of the fire, DOE officials had assured the public and firefighters that there was "no" elevated risk.[1] Meanwhile, a leading Hanford watchdog group demanded that the Environmental Protection Agency conduct a criminal investigation of DOE's failure to notify firefighters of an elevated risk.[2] A lack of clarity in talking about risk issues can intensify emotions surrounding events such as the 24 Command Fire.

Risk Assessment and Risk Perception

This episode reveals several important points about the term "risk" in regard to discussions concerning nuclear waste and cleanup. The waste and contamination legacy scattered around our nation's

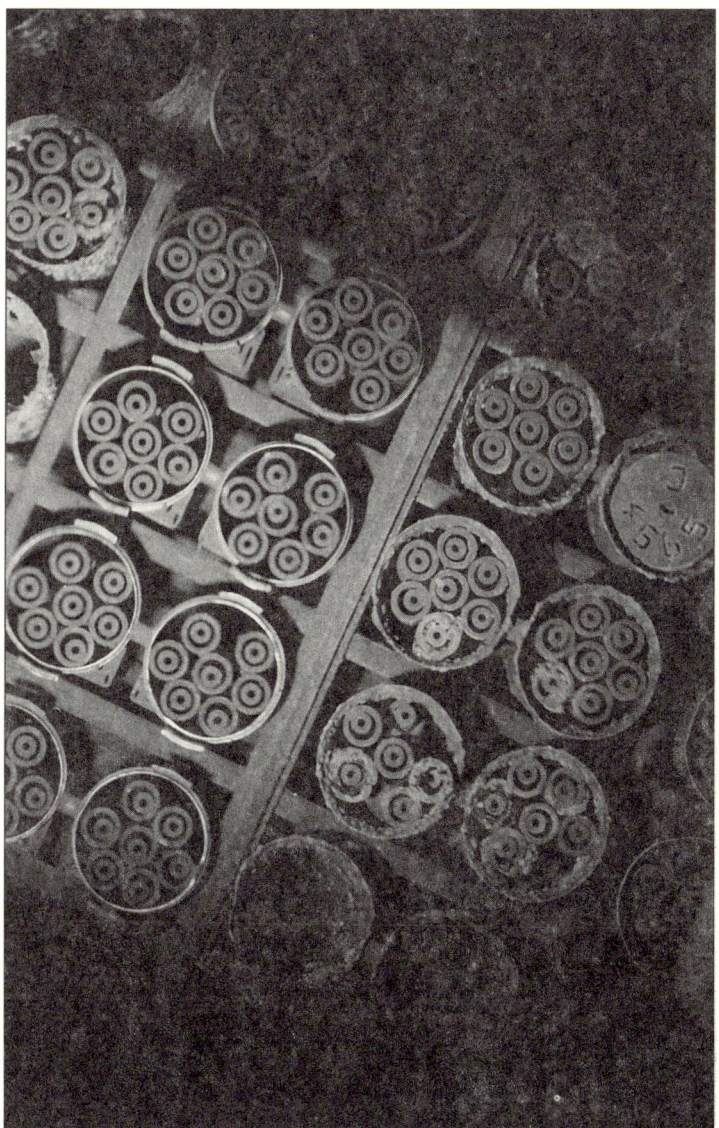

K-East Reactor fuel in wet storage, Hanford, December 31, 1992.

landscape indeed poses a degree of potential harm to surrounding populations. Also in 2000, large fires at the Idaho National Laboratory and at Los Alamos—especially at the latter—had real potential for releasing contaminants over wide areas.

In this chapter, I will focus on three aspects of risk assessment and perception that generate misunderstanding and mistrust between the technical community and lay citizens. These include:

- *Risk assessment and risk perception do not jibe.* Scientific risk assessment and the public's perception of risk often differ significantly. For many years, for example, scientists, engineers, and statisticians have all concluded, based on a large body of evidence, that smoking cigarettes is hundreds of times more harmful to public health than nuclear fall-out, and that driving a car is many times more dangerous than flying on a commercial jet. Just as often, however, public opinion surveys reveal very different perceptions about the relative risks posed by these activities.

- *People's gut reaction to involuntary and dread risk, not just to numbers and statistics, affects their choices.* Nuclear fallout, whether from weapons testing or from a nuclear facility fire, poses "involuntary" and "dread" (more correctly "dreaded") risk; potentially affected people do not choose this risk, whereas they do choose to smoke or drive. People are more likely to dread things nuclear because of making an association with nuclear war—images of Hiroshima, and from films such as *On the Beach*.[3] And, since the hazards posed by nuclear contamination are unseen and often latent, people are aware that they have no sensory clues to help protect themselves.

- *Scientifically-based standards for what is an allowable or unallowable risk usually do not take into account individuals' feelings about risk.* Scientists have used the extensive available data, including studies of Hiroshima's victims, to establish standards that define when exposures to radioactive isotopes pose an unacceptable health risk, and these standards are revised as more is learned. However, the standards are based on probabilities that some level of harm "will" occur to some unspecified individuals in a large population. An attempt to link cause and effect for any particular individual, however, is plagued by large uncertainties. So the process does not effectively address an individual's emotions associated with involuntary and dread risk.

Nuclear Facilities/Complexes Described in *America's Nuclear Wastelands*
(with site opening and primary operational dates)

Brookhaven National Laboratory (BNL): New York, 1947–

Fernald Site: Ohio, 1951–1989

Hanford Site: Washington, 1943–1989
   Pacific Northwest National Laboratory (PNNL), 1965–

Idaho National Laboratory (INL): Idaho, 1949–

Lawrence Livermore National Laboratory (LLNL): California, 1952–

Los Alamos National Laboratories (LANL): New Mexico, 1943–

Mound Facility: Ohio, 1948–1989

Nevada Test Site: Nevada, 1951–

Oak Ridge National Laboratory (ORNL): Tennessee, 1943–
   Joint Institute for Energy and Environment (JIEE): ORNL, TVA, and University of Tennessee, 1990–

Paducah Plant, Kentucky, 1952–

Rocky Flats Plant: Colorado, 1952–1988.

Sandia National Laboratories (SNL/NM): New Mexico/ California,1945–

Savannah River National Laboratory (SRNL): South Carolina, 1951–
   Defense Waste Processing Facility (DWPF), 1996–

Waste Isolation Pilot Plant (WIPP): New Mexico, 1982–

Weldon Spring Site: Missouri, 1955–1966

West Valley Site: New York, 1961–1980

Yucca Mountain Repository: Nevada, exploratory activities, 1987–

Here is one example of how involuntary and dread risk factors influence a person's perception. While the Hanford Environmental Dose Reconstruction Project was under way, we maintained a "hot line"—a toll-free number for people to call with any questions or concerns.

A caller from Whitman County, on the far eastern edge of Washington state, was quite concerned that the cancer suffered by a son-in-law might have resulted from Hanford's release of radiation into the environment in earlier years. As we talked, she said: "It's a good thing his wife has a good job with benefits at the bank. He had no health insurance after he lost his job at the fertilizer factory."

Based on the information at hand, in all likelihood this person had been exposed to relatively little nuclear contamination from Hanford, given his age and where he lived. I surmised—based on this admittedly quite limited conversation—that employment at a fertilizer factory was a more likely cause of the man's health problems. But, from the caller's point of view, the risks from working with fertilizer were familiar in a rural community, and accepted voluntarily when employed. Contamination that might have come from Hanford, on the other hand, was scary, involuntary, and mysterious.

Many experts who have studied risk perception and risk communication point out that perception *is* reality for many people.[4] The gaps between the conclusions of risk-assessing scientists and a broader public's perceptions have real consequences in the social and political arenas.

Regulatory bodies do not normally include involuntary and dread factors in their calculations when setting standards. However, most citizens do include these factors in their day-to-day decisions about which risks are acceptable. Most of us rely on an intuitive "gut feeling" when we make many choices about what to do or not do, and how to go about the business of living. People have visual images drawn from their own experiences or from various media sources that evoke "good" or "bad" reactions.

Of course, most of us also try to be "rational," and assess probabilities. How likely is a good or bad result if we choose one action over another? Thus, there is something of a scientist, too, in most of us. Still, gut feelings, or "affects" as social scientists call them, are likely to play a predominant role in our decision making. This appears to be quite true when it comes to making judgments about public or societal choices—such as nuclear power and nuclear waste disposal issues.

Some scientists and officials dealing with nuclear waste have defined risk as the probability of a release of contamination multiplied by its consequences. I would argue that people feel more strongly (have "affects") about the consequences of a release, rather than the prob-

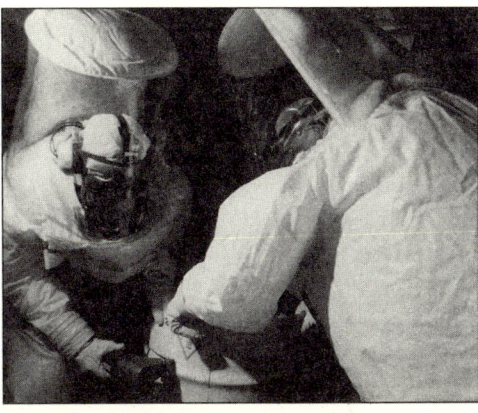

Decontamination and decommissioning progress, Rocky Flats Closure, September 24, 1998.

abilities that it will occur.[5] For many people, Murphy's Law—"If anything can go wrong, it will"—becomes a reasonable default position. (I will return to how consequences are assessed below.)

For a majority of Americans, the sense of dread from potential accidents at nuclear facilities or in transporting nuclear materials on highways is enhanced because of their having little experiential knowledge about the likelihood of such accidents. Contrast this, for example, with a driver making a decision to run a yellow light. Most people seem to regard the negative consequences of darting through a yellow light as relatively small, and estimate that the likelihood of incurring a fine from a policeman is very low, based on their experience and observation.

Involuntary Risk and a Sense of Impotence

The involuntary and dread aspects of risk perception exacerbate feelings of impotence. How can one protect one's life, loved ones, and values in a situation where arcane—and often inaccessible—experts decide what is safe?

Let me illustrate this by going back to my own early involvement with nuclear waste issues. In 1984, DOE's Office of Civilian

Radioactive Waste Management (OCRWM) released environmental assessments on nine potential sites across the nation for a national deep geologic repository where spent nuclear fuel and high level radioactive wastes would be disposed. Hanford was among the potential sites. In 1986, three finalists were selected for more intensive study, with Hanford among them.

As public hearings and meetings progressed, a chasm opened between the State of Washington and the OCRWM. Washington's political leaders, unlike those of other states, had not taken a "Hell, no!" stance. Instead, Governor Booth Gardner had said that Hanford might be an acceptable site if it were proven safe *and* if it was the safest site under consideration.

The Department of Energy's position was quite different. Any site would be acceptable, in its view, if it met risk-based standards set by the Nuclear Regulatory Commission (NRC) and the Environmental Protection Agency (EPA).[6] Its criterion was that a site must be "safe enough," not that it must be either "safe" or "the safest" among limited alternatives.

Theoretically, the rule-making process provided an opportunity for people and experts holding widely divergent views to engage in a dialogue about both societal and scientific factors to arrive at an acceptable set of standards or criteria. However, at the time of these public hearings, the NRC was still trying to design a process for its regulatory role, and the EPA had yet to propose its standards.

For those concerned about Hanford as a potential repository site, the frustrating result was a series of well-attended public hearings, where heart-felt expressions of concern were met with "comment noted" and assurances that DOE would be bound by as-yet-unadopted standards, set by agencies that were not officially present to hear comments.

How Risk-Based Standards Are Set

Probabilistic studies, based on biological laboratory work and epidemiological studies of previously exposed humans, form the basis for today's standards that regulate exposure to radioactive isotopes.[7] Incomplete knowledge from the past (as well as limits regarding our

Transuranic waste in storage at the Nevada Test Site.

understanding of circumstances in the future), however, generates significant degrees of uncertainty about how protective these standards are for any specific individual. Scientists must make assumptions about many complicated variables as they predict how protective the standards will be for the overall population.

In reconstructing past occurrences, as well as in predicting situations in potential future events, scientists must estimate the amounts of radioactive material released, how it reaches people, and how long it takes to do so. There are several pathways by which hazardous material can come in contact with people. Airborne particles or gaseous material may be inhaled or land on skin. Radioactive material also may settle on crops and orchards, so that people are exposed to contamination when consuming fruit or vegetables; cattle grazing on contaminated vegetation also can produce contaminated milk. Radioactive material in any form may contaminate surface water, or be washed into the soil and eventually reach groundwater. All of these pathways have probabilities and uncertainties associated with them.

Therefore, predictions about the damage done by a release of radioactive materials in a transportation accident, for example, would depend on assumptions regarding several factors:

- The nature of the material released: Different radioactive isotopes affect different organs of the body, and some are more penetrating than others. Some lodge in organs and may cause future problems; others pass through the body quickly. And, as noted earlier, the half-lives of radioactive materials vary.
- The number of people located close enough to be exposed directly, as well as the number who might be exposed if hazardous materials reached agricultural areas and streams, contaminating food and water.
- Weather and related conditions: Wind, rain, fire, etc.
- Situation of exposed people: Whether they are indoors or outdoors, old or young?

Estimates of health effects are grounded in past research and experience. Particularly, radiation effects are calculated based on linked dose-reconstruction and epidemiological studies. The most extensive of these deal with Hiroshima survivors; others include Pacific Islanders affected by early Pacific archipelago nuclear bomb tests, and people exposed to radiation from the Chernobyl disaster.

The science of this research is complex, but the basic idea is simple. The studies look for statistical relationships between heightened exposure to key radioactive isotopes and increased incidence of illness. For example, high exposure to radioactive iodine is associated with thyroid disease. Based on the knowledge accumulated from such studies, scientists can estimate the hypothetical increase in adverse health affects that would result from a significant radioactive release.[8] However, there are four important limitations:

- The epidemiological studies have not been conducted in controlled laboratory experiments—i.e., doses received and all the other potential factors affecting human health have not been studied under controlled circumstances. Thus, there is uncertainty about the estimates. A rule of thumb is that good dose estimates will be no more than three times larger, or three times smaller, than an actual dose.[9]

Fissile Materials

Nuclear weapons use highly "fissionable" materials, mostly plutonium, but also highly enriched uranium, which readily divide atomically to sustain chain reactions and quickly release incredibly-vast amounts of energy.

Both uranium and plutonium are toxic materials, and being heavy metals, they share many properties with lead, cadmium, etc. Both are low-energy/long-lived radioactive materials; once inhaled or ingested, they lodge in certain organs and emit a stream of alpha particles that can damage tissues. The half-life of Pu-239 is 24,110 years; the half-life of U-235 is 703,800,000 years.

Before enrichment processing, uranium-235 constitutes less than one percent of natural uranium, which is mostly uranium-238. After milling, uranium is enriched—the industrial process used to increase the proportion of U-235—in centrifuges that separate and concentrate U-235. Most commercial nuclear power reactors use "low enriched uranium" (LEU) containing 3% to 5% U-235. "Highly enriched uranium" (HEU) used in nuclear weapons, by contrast, contains more than 90% U-235. (Lesser "weapons-useable" amounts down to about 20% can produce crude, less-efficient bombs.)

The mission of producing "weapons-grade" HEU for the United States has been centered at Oak Ridge. In 1945, HEU fueled the Hiroshima atomic bomb.

The second material, plutonium-239, does not occur in nature. Plutonium is produced in nuclear reactors by bombarding uranium fuel with neutrons. The irradiated fuel is then treated chemically within separation plants—"canyons"—to separate the plutonium from unused uranium and other products of the nuclear reaction. There are several types of plutonium isotopes, but Pu-239 is considered weapons-grade.

Reactors and separations plants at Hanford, and later Savannah River, produced America's stock of Pu-239. During World War II, the Nagasaki bomb contained Hanford-produced Pu-239.

Stocks of U-235 and Pu-239 require strict supervision, control, and accountability. Relatively small quantities of either can provide the guts of a nuclear weapon. Therefore, the existence or development of nuclear facilities that enrich uranium or separate plutonium from irradiated reactor fuel (reprocessing) raise concerns about the potential proliferation of weapons-useable materials in the world today.

- Health effects are often latent, only appearing over periods of up to 30 years or more. Again, this increases predictive uncertainty, since other environmental conditions also may affect people's health over such a period.
- Partly as a result of the first two limitations, there is great uncertainty about the effects of "low-dose" radiation. The accepted scientific wisdom has been to extrapolate estimates about this, in a linear fashion, from the effects seen in studies linking higher doses to illness. There are, however, scientists who argue that low doses have no effect, or even beneficial effects, while others counter that low radiation doses have disproportionately larger effects on health over time.[10]
- When estimating potential effects and setting standards, the predicted relationships are averaged over a population. Scientists estimate the likely *increased rate of occurrence* of a disease among a hypothetical group of 1,000 (or 10,000, or 100,000) people. These are not projections about how any *particular* person may be affected.

Thus, on the one hand, this means that standards based on risk to health are averaged over large groups of people, often without taking into account the particular sensitivities that some individuals may have. On the other hand, however, standard-setters tend to be conservative—i.e., they set exposure limits lower than they otherwise might if they had more specific, less uncertain information.

To be clear, these standards for public exposure do not rely on calculations about the probability of an accident or release. However, regulations intended to prevent exposures that exceed these standards do rely on probability calculations about accidents. For example, rules governing the highway transport of high level nuclear waste are based on probability calculations of scenarios where containers would fail in accidents.[11]

In actuality, we have limited historical information available to help estimate the probability or magnitude of a widespread release of radioactivity—in the nuclear age, very few accidents at reactors have released radiation beyond their buildings, and even fewer transportation accidents have led to any release of radiation.[12]

What Is "Acceptable Risk"?

The basic proposition behind risk calculation is this: There is no risk-free activity. Everything one does entails some risk. Therefore, society must find ways to determine what levels of risk it will accept. An example in this regard: In the State of Washington, the Model Toxics Control Act, adopted as a citizens' initiative, requires the cleanup of hazardous waste sites wherever the risk of additional cancer incidence from a site is greater than 1 in 100,000.

Regulators often distinguish between risks to workers and those to a potentially affected population. The regulatory assumption is that workers, properly trained, are better able to protect themselves, and that they voluntarily accept any hazards associated with their employment. Generally, too, a workforce does not include children or the aged and infirm—those who would be more vulnerable to adverse effects from exposure to radiation or other contaminants.

Statisticians, economists, and risk assessors also often try to find some other societal benchmarks to help define what society collectively believes to be an "acceptable risk." Some turn to rates of occurrence of death or injury from voluntary activities. For instance, the rate of traffic fatalities per million persons in the United States, while gradually declining, is a relatively stable figure through time. Assume that this rate could be reduced still further by stricter laws, more law enforcement, greater investment in infrastructure, and enhanced safety devices in cars (at increased cost to owners); however, the public's choice not to support these measures represent society's acceptance of the existing fatality rate. Such understandings become benchmarks for researchers in deciding what the acceptable risks should be in other fields of human endeavor.

Another—very controversial—approach is to place some monetary value on human life. Such a value can be derived from the statistical analyses of court financial settlements and other data. Proponents of this approach argue that it makes sense to spend up to that amount to prevent a fatality, but not necessarily to spend more.

When stated this baldly, neither approach is likely to win wide acceptance. In part, this reflects my earlier point: Most people give more weight to the potential for catastrophic harm from involuntary, often-unseen exposures to environmental contaminants, than they do to admittedly risky but familiar activities.

In an ideal world, it would make sense to invest resources in those efforts that will save the most lives—many will argue, for example, that much of the approximately $6 billion a year spent on cleaning up nuclear weapons sites might be better diverted to breast cancer research or traffic safety improvements. However, given the many different societal values, the complex calculations about health, the uncertainty about predictions, and the efforts to balance present benefits against harm to future generations, such simple arguments appear to break down.

In summary, though DOE and regulatory officials rely on risk analysis and benchmarking to compare potential courses of actions and expenditures, society's conflicting perceptions of risk and the diverse values tied to those perceptions strongly influence cleanup and waste management decision-making in America today.

Notes

1. "Plutonium Spread Bad for DOE Credibility, if Not Public Health," *Seattle Post-Intelligencer*, August 6, 2000.
2. "Criticism Continues to Simmer in Wake of Hanford Fire," *Tri-City Herald*, August 20, 2000. "Official" assurances cited in the above newspaper articles were based on the following definition: Risk is the probability of the increased incidence of cancer within a large population as a result of exposure to radioactive or toxic material released into the environment.
3. For a fascinating in-depth discussion of images associated with nuclear activities, see Spencer R. Weart, *Nuclear Fear: A History of Images* (Cambridge: Harvard University Press, 1988).
4. Vincent Covello and Peter Sandman, "Risk Communication: Evolution and Revolution," in A. Wolhurst, ed., *Solutions to an Environment in Peril* (Baltimore: Johns Hopkins University Press, 2001), pp. 164–78.
5. There is some evidence that the manner in which probabilities are stated changes how people feel about them. People appear to be more risk averse in decision-making when told they have a 1 in 100 chance of a negative outcome, than when told the chance of a negative outcome is 1%. See Paul Slovic, "Risk as Analysis and Risk as Feelings: Some Thoughts about Affect, Reason, Risk and Rationality," *Risk Analysis*, Vol. 24, Number 2 (April 2004), pp. 311–22.
6. It is worth noting that neither regulatory body had adopted final standards at this point in time.

7. The discussion here focuses on radioactive contamination in order to be brief and clear. Note that calculations of risks and health effects from exposure to chemical contaminants essentially follow the same principles.

8. For a detailed discussion of dose reconstruction, see Committee on an Assessment of CDC Radiation Studies, *Radiation Dose Reconstruction for Epidemiologic Uses* (Washington, D.C.: National Academy Press, 1995).

9. The specific effects of a certain dose on a certain organ—the basis for dose conversion factors—is often supported by animal laboratory research. However, the uncertainties related to the actual doses received in an uncontrolled environment, as well as differences in the amounts that individuals might ingest, inhale, or absorb through the skin, all contribute to large variations in dose estimates.

10. For information on various approaches to estimating the health impacts of low radiation doses, see Richard Wilson, "Effects of Ionizing Radiation at Low Doses," American Association of Physics Teachers, *Resource Letter EIRLDP-1*, at the association's Web site: phys4.harvard.edu/~wilson/resource_letter.html. I am grateful to Dr. John Till for pointing out that, though significant uncertainties exist regarding low-dose effects, one would *not* expect them to have larger effects than those known with greater certainty to result from higher doses; correspondence with the author, October 6, 2006.

11. See Chapter 7.

12. Three Mile Island and Chernobyl are the best-known reactor accidents; other significant nuclear waste handling and processing accidents also occurred in the former Soviet Union, but the information linking radiation releases to health effects is limited.

Disposal silos unearthed during long-term remediation efforts by DOE's Office of Environmental Management, Hanford, February 28, 2005.

3

The Legal and Regulatory Basis for Cleanup

This chapter describes the regulatory framework in which risk-based standards are developed, and the institutional structure in regard to how it is applied to nuclear facilities. Three key historical and legal facts shape cleanup activities at our nation's nuclear sites:

- The nuclear age started in wartime, and expanded during the early years of the Cold War. Therefore, the federal government dominates in the laws and regulations specific to nuclear-related health and safety.
- Other environmental laws, however, provide for significant State roles in regulating hazardous substances to protect public health and safety.
- The Department of Energy and its predecessors—the Army Corps of Engineers, Atomic Energy Commission, and the Energy Research and Development Administration—have been largely self-regulating in the operation of nuclear facilities.

Federal Pre-emption

The U.S. Supreme Court has long held that federal laws "preempt" state laws in areas where Congress has appropriately "occupied the field." Congressional authority to do so is rooted in various Constitutional provisions, including the "commerce clause" and the "supremacy" clause.[1]

Damaged N-Reactor fuel element, Hanford, March 31, 1996.

Citing this general principle, federal courts have held that Congress preempted the field of nuclear health and safety regulation through the passage of the Atomic Energy Act (1946) and the Energy Reorganization Act (1974). Consequently, the courts generally have struck down state laws that attempt to regulate specific processes or impacts of defense nuclear activities.[2] States do have laws and regulations that apply to civilian nuclear reactors and the disposal of wastes from these facilities. However, such laws and regulations generally must comply with, or be equivalent to, regulations promulgated by the Nuclear Regulatory Commission (NRC).[3] Similarly, state laws and regulations concerning the transportation of nuclear wastes and materials must be consistent with federal regulations.

As public concern mounted in the 1980s and 1990s over the residual waste and contamination problems at defense nuclear facilities, state political leaders looked for leverage. For the most part, they had to find it elsewhere than in adopting laws directly aimed at nuclear activities. The pressure to act, however, was strong—driven by the public's distrust of self-regulation by the historically secretive and insulated Department of Energy.

The States' Role in Environmental Law

The basic model used in federal environmental law is this: The federal government adopts basic standards and enforcement provisions, and then approves and financially supports state programs that adopt and enforce standards that are at least as stringent as the federal ones. In essence, the Clean Air Act (1963, 1970), Clean Water Act (1972, 1977), and the Resource Conservation and Recovery Act (1976; regarding the handling and disposal of hazardous chemicals) all follow this model. Congress has usually directed (or assumed) that federal agencies and facilities are subject to state regulation under these acts. Indeed, in the mid-1980s, states began to use these laws to participate in the regulation of nuclear facilities.

In 1986, a federal district court decided in favor of Tennessee's effort to regulate wastewater discharges at Oak Ridge. DOE declined to appeal. By 1989, several states were pursuing regulation through the Resource Conservation and Recovery Act as well as the Clean Water Act.

Public Distrust

As the Cold War ebbed during the late 1980s, watchdog organizations and the Department of Energy itself made revelations that raised concern and distrust among many citizens. The DOE, under pressure from the news media and watchdog groups in Washington state, released 19,000 pages of previously classified documents, showing that early off-site radiation releases from Hanford had been far more significant than the federal government previously acknowledged. At Rocky Flats, Colorado, the FBI began a criminal investigation of potential violations of environmental laws. In Idaho, political leaders drew attention to poorly-disposed wastes threatening the Snake River Plain Aquifer, a major water source for the agricultural sector.

A number of Congressional committee hearings also were focusing on specific problems and lapses of safety procedures at defense nuclear sites. The Department of Energy, meanwhile, was releasing a great deal of information in environmental assessments required by the Nuclear Waste Policy Act (1982, 1987) in regard to siting a national deep-geologic repository, as well as through specific environmental impact statements at several sites, including Hanford.

In 1989, Admiral James Watkins, the incoming Secretary of Energy, promised a radical shake-up of the culture at DOE. He sent well-

Chronology of Key Congressional Legislation and Presidential Executive Orders

Manhattan Engineering District (MED) funding—1941 to 1946

Atomic Energy Act—1946 (1954)

Clean Air Act(s) and Amendments—1963 (1967, 1970, 1977, 1990)

Freedom of Information Act (FOIA)—1966 (amended 2002)

National Environmental Policy Act (NEPA)—1970

Clean Water Act (CWA)—1972 (amended 1977, 1978)

Energy Reorganization Act—1974

Resource Conservation and Recovery Act (RCRA)—1976

DOE Organization Act—1977

Uranium Mill Tailing Remediation Control Act (UMTRCA)—1978

Comprehensive Environmental Response, Compensation and Liability Act (CERCLA)/"Superfund"—1980

Low Level Radioactive Waste Policy Act—1980 (amended 1985)

West Valley Demonstration Project Act—1980

Nuclear Waste Policy Act (NWPA)—1982 (amended 1987)

Federal Facilities Compliance Act—1992

Waste Isolation Pilot Plant Land Withdrawal Act—1992

Executive Order 13084, Consultation and Coordination with Indian Tribal Governments—1998

Energy Employees Occupational Illness Compensation Program Act (EEOICPA)—2001

Homeland Security Act/Critical Infrastructure Information (CII)—2002

Executive Order 13392, Improving Agency Disclosure of Information—2005 publicized "tiger teams" to review procedures at the department's various nuclear sites, and he established the Office of Environmental Management (EM) program within DOE.

The result of these and related developments was paradoxical. For the first time, the Department of Energy's nuclear operations came to the notice of a relatively broad, cognizant public. "Stakeholders" in what happened at the sites emerged outside the department, its contractors, and the communities directly dependent on DOE jobs. There was considerable media coverage, and the department itself was far more open than in earlier years.

The newly assembled public, aroused by the revelations, was very mistrustful. They wanted independent oversight and regulation—usually state or local—of DOE's activities. In the fall of 1989, for example, the media in Washington state picked up on the fact that some of Hanford's older, high level radioactive waste tanks contained quantities of ferrocyanide. This compound, which had been used to precipitate solids in earlier waste management days, was potentially explosive above certain temperatures. Media reports noted that residual radioactive materials in the tanks were currently generating heat and raised the specter of a tank explosion due to the presence of ferrocyanide.

At the time, I was part of the lightly-staffed state office then evolving to oversee Hanford's waste management. Our leading tank experts were attending meetings outside the state. Christine Gregoire, director of the Ecology Department, called and ordered me to go to Hanford and investigate the situation immediately. Her comment: "Mike Lawrence [then manager of DOE's Richland Operations Office] has told me that everything's under control. But I can't tell the press that it's okay because Mike says it is. We need an independent look." Thus, I visited Hanford and developed an independent assessment of the situation for the director.

Outside pressure on our office to "control" DOE's activities were constant. Similar media and public pressures developed in Ohio, Colorado, Idaho, Tennessee, Nevada, New Mexico, and South Carolina as well.

Earlier that same year, the State of Washington, DOE, and the federal Environmental Protection Agency had negotiated the "Hanford

Federal Facility Agreement and Consent Order"—or "Tri-Party Agreement"—that provided for the state regulation of hazardous waste management at Hanford.[4] Several other states also quickly followed with compliance or oversight agreements at defense nuclear sites.

Sovereign Immunity and the Federal Facilities Compliance Act

As noted earlier, the states could not directly regulate nuclear activities on the basis of "radiological" health and safety impacts. Therefore, agreements, such as that at Hanford, relied on the fact that much of the waste at DOE sites was "mixed"—i.e., it contained both radioactive and hazardous "chemical" components. The states could regulate the latter under authority delegated by the EPA pursuant to the Resource Conservation and Recovery Act (1976).

However, Congress had not been clear enough about its intent regarding the states' regulatory enforcement authority over federal facilities. When the states began to assert such authority, federal attorneys resisted, based on "sovereign immunity." This doctrine carries over from English common law the notion that the crown (or government) cannot be sued. Congress, however, can waive sovereign immunity, and it did so, in the Federal Facilities Compliance Act (1992).

The states mentioned in the section above, and others, worked together to secure this change in the law. Subsequently, federal courts have upheld the authority of the states to regulate the management of mixed hazardous chemical and radioactive wastes.[5]

Comprehensive Environmental Response, Compensation and Liability Act (CERCLA)

The Resource Conservation and Recovery Act (1976) deals with "active" chemical management and waste operations, whereas the "cleanup" of sites where hazardous wastes were released into the environment in the past is the province of the Comprehensive Environmental Response, Compensation and Liability Act (1980). This superfund law differs from RCRA (and clean air and water laws) in three significant respects when it comes to cleaning up federal nuclear sites.

First, CERCLA does not contain a delegation of authority to the states. It is administered by the Environmental Protection Agency. Second, the law is structured to get the "potentially liable parties"—those who contributed to the problem—to clean it up or, failing that, to reimburse the government for cleaning up. Third, because CERCLA is administered by a federal agency, its regulations can directly address the hazards of radiation.

A crew recovers PCBs (Polychlorinated Biphenyls) in a gulch at the Los Alamos National Laboratories, September 30, 2000. This "chemical" pollution resulted from the storage of electrical transformers in the area.

In practice, the application of CERCLA at federal nuclear sites has varied. DOE, as the liable party, has been the key decision-maker at some sites, whereas the EPA's management and direction of cleanup has been more prominent at others. Some federal facility agreements, such as those at Hanford and Rocky Flats, contain provisions aimed at integrating state RCRA authorities and CERCLA decisions. Others, such as that covering Weldon Spring, Missouri, near St. Louis, relegate states to an advisory role in CERCLA decisions.

In any case, CERCLA does provide that "applicable, relevant and appropriate" state regulations and standards be taken into account. As already noted, however, state standards specific to protecting public health from nuclear contamination are subject to legal challenge.

One significant CERCLA action at Hanford illustrates these complexities. More than 80% of DOE's stock of spent nuclear fuel was stored in water basins near the Columbia River at two shut-down reactors. One of the basins had leaked. Both presented serious threats in the event of an earthquake, according to a report issued by the Defense Nuclear Facilities Safety Board in 1994.[6] DOE gave high priority to retrieving the spent fuel from the water basins, and stabilizing and storing it safely away from the river.

As the 1990s wore on, however, technical issues and cost overruns plagued the program. The state and regional EPA officials concluded that some kind of enforcement action was needed to get the project back on track. As the major threat was the release of radiation to the atmosphere or the river, these officials decided that EPA should take the lead. That agency has the authority to order a "removal action" under CERCLA, where a leak of hazardous waste has occurred. While the water basins also were active waste management units, the state realized that its efforts to enforce cleanup under RCRA would likely be challenged legally on the grounds of federal preemption in the area of nuclear safety. Such litigation would lead to further delays.

For the most part, CERCLA cleanup decisions—called "remedy selection"—revolve around the issues mentioned in Chapter 1:

- Should contamination be left in place and provided with caps and barriers to prevent its movement through the environment, or should it be removed ("muck and truck") to an engineered disposal facility?
- What is a safe (or acceptable) residual level of contamination that can be left behind?

CERCLA provides that these decisions at a given site will be reviewed every five years to see if the results meet expectations. If not, decisions can be changed.

Problems of Scale and Complexity

The example of Hanford's spent fuel storage is just one illustration of the scale and complexity of the cleanup at many defense nuclear sites. The drafters of RCRA and CERCLA contemplated cleanup occurring at single factory or landfill sites. They did not focus on such facilities as Hanford, Rocky Flats, Los Alamos, the Idaho National Laboratory, or the Nevada Test Site where a combination of old disposal sites leaking contaminants, complexes managing highly dangerous wastes, and currently active nuclear facilities were all intermingled. At many of these locations, such co-mingling is scattered across tens to hundreds of square miles.

At Hanford, for example, there were more than 1,500 sites where releases of contaminants to the environment had occurred from 1943 to 1989. There also were 80 or more active waste management facilities, with many of the latter located on top of the early release sites; contamination from both were intermixed in the soil and groundwater.

To DOE and its contractors—who had been self-regulating for nearly a half-century—the imposition of both CERCLA and RCRA rules and authority often appeared to be redundant, costly, and unnecessary. To the public and Congress, who were suspicious of DOE and concerned about the spread of contamination, the exercise of this authority was essential. They expected the states, EPA, and DOE to figure out how to make the regulatory regime work.

Nuclear Waste Definitions

In my experience, beginning a discussion with the definitions of nuclear waste contributes more confusion than enlightenment. Therefore, I have left a brief explanation of these definitions to this relatively late point in the text. As I shall suggest shortly, the principal importance of these definitions is geographic—i.e., *where* waste will be disposed of is a function of the definitions.

In the United States, most nuclear wastes fall into three categories:[7]

- *High level waste.* High level waste is defined by source or origin, not, as the name implies, by concentrations of radionuclides. This is waste resulting from the dissolution of spent nuclear fuel to extract plutonium or uranium. Under the Nuclear Waste Policy Act of 1982, spent nuclear power plant fuel also is included in this category.
- *Transuranic waste.* Transuranic waste contains concentrations of radioactive isotopes with atomic numbers greater than uranium[8]

(thus, "trans" uranic), including various isotopes of plutonium, americium, neptunium, etc. This is a concentration-based definition. Wastes containing more than a certain amount of these transuranic elements per gram are classed as transuranic. These isotopes do not emit very penetrating energy, but have extremely long half-lives, during which they remain toxic.

- *Low level waste.* Low level waste is, essentially, everything else.[9] Low level waste ranges from clothing and tools potentially contaminated by being used in radiation facilities, to the extremely "hot" internal equipment from nuclear reactors. Therefore, low level waste is divided into three subcategories—Classes A, B, and C—depending on the concentrations of various radioactive elements. Class C wastes may be as dangerous as high level waste.

Figure 3.1 provides a more complete set of definitions, as summarized by the Oregon Department of Energy's Nuclear Safety Division. These definitions were set by the Atomic Energy Commission relatively early in its existence, and govern civilian nuclear waste activities and, to a degree, defense nuclear activities as well. However, the elimination of the Atomic Energy Commission and the separation of its former functions under the Energy Reorganization Act (1974) left it to the new Department of Energy (established 1977) to regulate its own defense nuclear activities, and its definitions now vary slightly from those carried on by the Nuclear Regulatory Commission (established 1974) in the civilian sphere.

It also is worth noting that regulations covering the transportation of nuclear waste are based on a different set of concentration-based categories. We shall return to transportation issues in a subsequent chapter.

For our purposes here, the significance of the three-pronged definitional scheme for waste is this: Congress has determined by law that each of the three categories of waste will be disposed of differently:

- *High level waste and spent nuclear fuel* is to be disposed of in a deep geologic repository at Yucca Mountain, Nevada. Therefore, these wastes are not to remain in place at sites, but eventually will be removed and transported to the Nevada repository (scheduled for opening in 2019).

Figure 3.1
Nuclear Waste Definitions

Type of Waste or Material	Definition/description*
Spent Fuel	Spent or used fuel elements from nuclear reactors. The spent fuel is highly radioactive and must be stored in special facilities that shield and cool the material. In the United States, the spent fuel is primarily from commercial nuclear power plants.
High Level Waste	Material generated by the chemical reprocessing of spent fuel. The waste is highly radioactive and must be isolated from the environment for thousands of years.
Transuranic Waste	Waste generated primarily during the research, development, and production of nuclear weapons. Most of the waste consists of such things as laboratory clothing, tools, gloveboxes, rubber gloves, and air filters, contaminated with small amounts of plutonium and other radioactive elements. Some of these will remain radioactive for tens of thousands of years.
Low Level Radioactive Waste	Any radioactive waste that does not fall into one of the other categories. Most low-level waste contains small amounts of radioactivity in large volumes of materials. Some low-level waste, however, can contain significant levels of radioactivity. The Nuclear Regulatory Commission classifies low-level waste into four categories according to the level of hazard. Low-level waste that can be disposed of by shallow land burial is classified as "A," "B," or "C"—from least to most hazardous. Low-level waste that is too hazardous for shallow land burial is called "Greater than Class C" waste. There is currently no disposal facility for this waste.
Low Level Mixed Waste	Waste that contains both radioactive and chemically hazardous materials. The radioactive component of mixed low-level waste is similar to the component of low-level waste.
Low-Activity Waste	"Low-activity" waste is not a regulatory term. It refers to the less-radioactive waste that remains following the process of separating the highly radioactive constituents from high-level waste. While sometimes used interchangeably with the term "low-level" waste, low-activity waste is not the same.
NORM	Naturally Occurring Radioactive Material (NORM) includes radium, radon and other radioactive elements that exist in the earth's crust. Some NORM has a relatively high concentration in a small volume—such as industrial gauges. Most NORM has a low concentration of radioactivity in a large volume.

* The definitions provided here are not intended as comprehensive technical definitions. They are provided to help give a basic understanding of each type of waste/material.
Source: Oregon Department of Energy, Nuclear Safety Division.

- *Defense-related transuranic wastes* are disposed at the Waste Isolation Pilot Plant, in operation outside of Carlsbad, New Mexico. Again, this means that waste will be removed from where it originally was generated and stored.
- *Low level waste* may be disposed of at the sites where generated, or moved to regional disposal sites. The regional pattern differs depending on whether the low level wastes in question result from defense or civilian nuclear activities.

Notes

1. "State law can be preempted in either of two general ways. If Congress evidences an intent to occupy a given field, any state law falling within that field is preempted. If Congress has not entirely displaced state regulation over the matter in question, state law is still preempted to the extent it actually conflicts with federal law, that is, when it is impossible to comply with both state and federal law, or where the state law stands as an obstacle to the accomplishment of the full purposes and objectives of Congress."—*Silkwood v. Kerr-McGee Corp.*, 464 U.S. 238, 250 (1984).
2. "The Supreme Court has repeatedly held that the Atomic Energy Act occupies the field with respect to these radioactive materials [covered by the Atomic Energy Act] for safety purposes, and state laws in that field are thus preempted."—Department of Justice brief seeking summary judgment against the State of Washington's initiative to regulate the Hanford cleanup, citing *Silkwood*.
3. Federal law applies specifically to materials used in, or produced by, nuclear reactors. States generally have broad authority to regulate "naturally occurring" and accelerator produced radioactive materials, such as radium and small quantities of materials used for medical diagnoses and treatments.
4. See a summary in Roy E. Gephart, *Hanford: A Conversation about Nuclear Waste and Cleanup* (Columbus: Battelle Press, 2003), ch. 7. This agreement combined state authorities under the Resource Conservation and Recovery Act (1976) with EPA authorities under the Comprehensive Environmental Response, Compensation and Liability Act (1980). The agreement also contained provisions for amendment and dispute resolutions. The parties have agreed to about 300 amendments since 1989. For more information, see www.hanford.gov/?page=92&parent=90.
5. For example, while ruling that a recent Washington state initiative to regulate waste disposal at Hanford was preempted because it attempted to regulate radionuclides, U.S. District Judge Alan McDonald cited several cases making it clear that the states have authority to regulate the hazardous non-radioactive components of mixed waste. See McDonald, "Order Granting Motion for Summary Judgment" in *United States v. Manning*, CV-04-5128-AAM.
6. Defense Nuclear Facilities Safety Board, "Recommendation 94-1 to the Secretary of Energy," May 24, 1994.
7. There is no single, clear statement of definitions in the law and regulations, and definitions have evolved under the authority of several federal agencies. Definitions may be

found in 10 CFR 60 ff (Nuclear Regulatory Commission), 49 CFR 173 (Environmental Protection Agency), 40 CFR 60 ff (Department of Transportation), and USDOE Order 435.1. The table of definitions created by the Nuclear Safety Division of the Oregon Department of Energy included in the text provides a good summary.

8. The atomic number of uranium is 92.

9. There are two exceptions that will not be pursued here. First, naturally-occurring and accelerator-produced radioactive materials (NORM/NARM) are outside the definitions of the Atomic Energy Act and are, contrary to the earlier discussion, subject to state regulation. This includes, among other things, large numbers of isotopes used for medical diagnosis and treatment. Second, "Greater than Class C" waste consists mainly of radium sources (naturally-occurring) used in testing and measurement gauges. The latter are extremely dangerous, subject to both federal and state regulation, and a source of concern about diversion for terrorist purposes.

AEC and Its Successors

Atomic Energy Commission (AEC)—founded 1946; assumed control of Manhattan Project facilities plus new functions over 28 years of operations. Congress separated the **AEC** into the **NRC** and **ERDA** (see below) with the Energy Reorganization Act of 1974.

Nuclear Regulatory Commission (NRC)—founded 1974; successor to **AEC**. Conducts oversight/safety/licensing regarding commercial-electric nuclear power plants, medical/industrial/academic nuclear materials, storage/transport/disposal of nuclear waste, and decommissioning of nuclear facilities.

Energy Research and Development Administration (ERDA)—founded 1974; successor to **AEC**; incorporated into **DOE** in 1977. From 1974 to 1977, undertook functions not assumed by **NRC**. **ERDA** oversaw nuclear weapons production and management, naval nuclear propulsion, waste management, and energy development and conservation programs.

U.S. Department of Energy (DOE, or USDOE)—successor to **AEC**; established by DOE Organization Act of 1977, largely in response to the oil crisis of the early 1970s. Combined **ERDA** and other energy related federal programs in development/regulation/research regarding coal, petroleum, natural gas, electric, renewable, alternative, and nuclear defense programs. **DOE** agencies involved with defense nuclear waste issues include the following:*

National Nuclear Security Administration (NNSA)
Office of Civilian Radioactive Waste Management (OCRWM)
Office of Environmental Management (EM)
Office of Health, Safety, and Security
Office of Legacy Management (LM)
Office of Nuclear Energy
Office of Science
Office of Secure Transportation (OST)

* It is essential to note that within DOE, different secretarial program offices "own" different sites. Some examples—the Office of Environmental Management (EM) "owns" most of Hanford; however, the Office of Science "owns" Hanford's Pacific Northwest National Laboratory. On the other hand, the Office of Nuclear Energy "owns" the Idaho National Laboratory, though EM also conducts major cleanup there. At Oak Ridge, such arrangements are even more intricate.

In regard to DOE's National Nuclear Security Administration (NNSA; established 2000 regarding security/nonproliferation, nuclear weapons safety/reliability, and naval nuclear power), it "owns" several major programs including supervision of the weapons labs (e.g., at Los Alamos and Lawrence Livermore) and is somewhat insulated from management by the Secretary of Energy.

The Office of Legacy Management (LM) has inherited Rocky Flats and Fernald after the Office of Environmental Management's closure programs at those sites.

This complexity of organization can make transparency, disclosure, and public accountability difficult to achieve.

Other U.S. Government Departments and Agencies Participating in Nuclear Waste Disposal and Stewardship Programs

Office of Management and Budget (OMB)—within the Executive Office of the President of the United States

Tennessee Valley Authority (TVA)

U.S. Army Corps of Engineers (USACE)

U.S. Bureau of Land Management (BLM)

U.S. Department of Defense (DOD)

U.S. Department of Health and Human Services (HHS)
Agency for Toxic Substances and Disease Registry (ATSDR)
Centers for Disease Control
National Institute for Occupational Safety and Health (NIOSH)

U.S. Department of Homeland Security (DHS)
U.S. Coast Guard (USCG)

U.S. Department of Transportation (DOT)
Federal Highway Administration (FHA)
Federal Railroad Administration (FRA)

U.S. Environmental Protection Agency (EPA)

U.S. Fish and Wildlife Service (FWS)

U.S. Government Accountability Office (GAO)

4

Politics, Jobs, and Public Engagement

Nuclear Site Politics

Political considerations were not a primary part of the locating process in selecting the Manhattan Project's original nuclear sites, other than at Oak Ridge (as described in Chapter 1). In fact, there appears to have been little consultation with state and local politicians over the selection of Los Alamos and Hanford. Once the genie was out of the bottle, however, maintaining or expanding these sites and locating new ones, such as at Savannah River, South Carolina, and Rocky Flats, Colorado, became the stuff of Congressional logrolling.

When I first became involved with Hanford, located near the City of Richland, the site employed over 18,000 workers. Union jobs were paid at prevailing federal (Davis-Bacon) wages. Benton County, in which the Hanford Site is situated, had the highest education levels and the highest per capita number of Ph.D.s of any county in Washington state.

Richland, Oak Ridge, and Los Alamos were federal company towns, until amendments to the Atomic Energy Act in the1950s provided for their transfer to private ownership and local government. At the end of the 1980s, a newly-installed city official in Richland, whose previous experience included working for the City of Oak Ridge, described the situation this way: "These are like colonial towns. The Department of Energy comes in, pumps money and jobs into the local economy, and expects the locals to be grateful. Field office officials are supposed to keep the locals in line in return for the economic

East Tennessee Technology Park in the Oak Ridge complex, January 1, 1993.

benefits." This particular official had quickly found himself on a major Hanford contractor's "unreliable" list because he had become an independent advocate for local interests.

By virtue of origins in the Manhattan Project and the Cold War that followed, the communities situated around the larger defense nuclear facilities were isolated and insulated. No one on the outside was supposed to know what was going on at these sites. Those inside interacted mostly with one another insofar as work, education, and ideas were concerned. State officials were supportive of the economic benefits, and were under relatively little political pressure to oversee the sites because most of the state-wide electorates knew little about them.

The concentration of nuclear scientists and engineers, as well as skilled labor, made these communities attractive places for other non-defense activities. Congress amended the Atomic Energy Act in 1954 in order that land and facilities at Atomic Energy Commission sites could be leased for other "nuclear-related" purposes. At Hanford, this resulted in a commercial, low-level radioactive waste disposal site, a nuclear fuel fabrication facility, and three planned public power reactors (two of which were never finished, largely due to eventual statewide public opposition and high cost considerations). A complex of commercial nuclear waste treatment facilities developed around Oak Ridge, too.

Also at Hanford, under the aegis of two powerful U.S. senators (Warren Magnuson and Henry M. Jackson), a steam-turbine electric generating plant was attached to the N-Reactor to feed power to the Bonneville Power Administration grid, thus augmenting the existing regional public hydro-electric system. Later, when national targets for plutonium production had been met, these Washington senators arranged for the N-Reactor to continue operating in order to produce electricity. (A large portion of the spent reactor fuel stored in the K-Basins near the Columbia River resulted from this arrangement. This fuel was not primarily irradiated for the weapons-grade plutonium that could be extracted from it.[1])

From about the mid-1970s, however, a series of factors eventually would change the relatively cozy "live and let live" character of local and regional politics in regard to defense nuclear sites:

- Waste wars—i.e., a growing public and media concern over nuclear waste disposal issues, spurred in part by the attention given to the federal government's efforts to locate new facilities, the failures of commercial nuclear waste operations in Kentucky and Illinois, and the Carter administration's decision to move away from reprocessing spent nuclear fuel.
- Growing public resistance to new civilian nuclear power plants in the wake of the Three Mile Island and Chernobyl accidents.
- Organized public opposition to the Reagan administration's accelerated nuclear weapons production effort.
- Decline and final collapse of the Soviet Union in 1991.

A Broader Public Engages

In 1974, the United States appeared to be poised on the brink of a major expansion of civilian nuclear power. When I attended an

American Planning Association conference in Atlanta that spring, a plenary speaker advised that the country must be ready to site an additional 100 nuclear power plants by the end of the decade. In that same year, Congress divided this role out from the Atomic Energy Commission and created the Nuclear Regulatory Commission to deal with the coming commercial expansion.

In the same time period, the Atomic Energy Commission (and its successors, the Energy Research and Development Administration, 1974, and then the Department of Energy, 1977) had accelerated the search for sites to dispose of spent fuel, and high level and transuranic wastes. As early as 1957, the National Academy of Sciences had suggested using deep geologic disposal for long-lived and dangerous radioactive wastes. In 1970, scientists further suggested disposal in underground salt formations. Bedded salt's appeal is two-fold: Geologically it is extremely stable, remaining in place for hundreds of millions of years, and salt formations are self-healing when breached.

From 1970 to 1972, the AEC explored an abandoned salt mine near Lyons, Kansas, but it then abandoned the project after local residents and officials raised many concerns about holes drilled through the

Lead reactor parts unearthed during remediation of Hanford's reactor areas along the Columbia River, February 28, 2005. The contaminated parts had been disposed in this burial ground during reactor operations.

salt formation. This was among the earliest cases of where the "siting" of a waste disposal facility aroused broad public concern.

By 1977, waste disposal issues were gaining ever greater attention. Efforts to reprocess spent commercial reactor fuel at Morris, Illinois, and West Valley, New York, had run into serious technical difficulties and escalating costs. Reprocessing—the recycling of usable uranium from the irradiated fuel—indeed would have produced a relatively small volume of liquid high level waste, when compared to the growing amount of spent fuel from nuclear plants then operating across the nation. But reprocessing also provided the opportunity to extract plutonium, which was created during the initial burning of the fuel, as well as "enriched"[2] isotopes of uranium. Consequently, nuclear weapons non-proliferation issues arose. The federal government also had supported research into the "breeder reactor," which would capitalize on the recycling of uranium and plutonium, thus reducing the need to mine and refine uranium. However, President Jimmy Carter decided—as a matter of international nuclear non-proliferation policy—that the federal government no longer would support reprocessing and research into breeder reactors.

In the wake of the Three Mile Island accident (1979), public anti-nuclear organizations grew and focused on the absence of a safe national waste disposal plan as a major reason to oppose the construction of new nuclear power plants. Thus, questions regarding spent fuel disposal ascended in the national political agenda. At the same time, low level radioactive waste disposal operations also captured more public attention. It was revealed that commercially operated disposal sites in Illinois and Kentucky had not isolated contaminants from the surrounding soil and groundwater. The sites were closed, and, due to the site operators' insufficient resources, state officials faced significant cleanup costs. In Washington state—the location of one of only three remaining disposal sites across the country—voters approved a "Don't Waste Washington" initiative, banning out-of-state wastes by a three-to-one margin in November 1980.

In that same year, Congress adopted the Low Level Radioactive Waste Policy Act, which provided that groups of states could form regional interstate compacts to regulate the movement of radioactive wastes. (In Chapter 6, I will return in more detail to the interstate and

interregional political struggles around the siting of waste facilities.)
Here, the important point is that nuclear waste disposal—as well as
the siting of new nuclear plants—mobilized many people not previously interested in nuclear policy issues.

Repository Siting and Weapons Rearming in the 1980s

At the end of 1982, Congress passed the Nuclear Waste Policy Act
(NWPA), aimed at siting deep geologic repositories for spent nuclear
fuel and high level waste. President Ronald Reagan had made it
clear that he had no objection to spent fuel reprocessing—so long
as it was economically viable without government help. It proved
not to be, however. Thus, the issue of spent fuel disposal remained
a major hurdle for those who wanted to expand nuclear power in
the United States. The NWPA accepted disposal as a federal responsibility, but arranged that it should be paid for by a surcharge on
nuclear power.

The NWPA included many complex provisions. Two, in particular,
increased opportunities for public participation. First, the legislation
provided that states would be involved in the assessment of potential
repository sites and would have a conditional veto. Second, the act
provided that the President could decide to "co-mingle" defense high
level waste with civilian spent fuel and high level waste in the same
repositories. The Reagan administration opted for co-mingling in
1985.[3]

The NWPA anticipated two repositories. The first—a western or
southern—would be located in either basalt (Hanford), welded tuff
(Yucca Mountain, Nevada), or salt formations in Texas, Louisiana,
or Mississippi. The second would be situated in granite formations
in the Carolinas, upper New England, or the upper Midwest. As a
result, public concern over nuclear waste now spread to new areas,
such as Maine, Vermont, North Carolina, Michigan, Wisconsin,
and Louisiana, where the issue previously had not attracted much
interest, and where defense nuclear facilities policy had little prior
impact. DOE also published and released information on the two
defense sites under consideration—Hanford and the Nevada Test
Site—where up to this time little information about either facility

had ever been made available to the public. In all of these cases, the states were subsidized to hire staff or consultants to review information and to engage the public.

The act imposed an unrealistic schedule for the siting and development of the national repositories. That schedule, though, forced the Department of Energy to mobilize quickly to meet the public in ways it was quite unaccustomed to doing.

In the summer of 1984, the publication of environmental assessments regarding each of the nine sites launched a series of emotionally heated public hearings in all of the involved states and regions. In the spring of 1986, DOE narrowed the search for a first repository to three sites (Hanford, the Nevada Test Site, and Deaf Smith County in the Texas Panhandle), and suspended the search for a second repository. Many asserted that the latter decision was intended to remove the issue from Senatorial politics in a number of the "granite" states.

In these same years, the Reagan administration was stepping up production of nuclear weapons as part of an increasingly hard-line policy toward the Soviet Union. A number of anti-nuclear organizations expanded in the face of this build-up, including the Committee for a Sane Nuclear Policy (SANE) and Nuclear Freeze. There were repeated anti-nuclear demonstrations, including civil disobedience, at Rocky Flats and the Nevada Test Site.

The pace of weapons production at Hanford had slowed dramatically from a high point in the 1960s, when the N-Reactor was completed, and the Kennedy administration's initiative to "close the missile gap" had put Hanford's older reactors running full tilt. In the early 1980s, however, the PUREX plant (for recovering uranium and plutonium from used nuclear fuel)—the most recent of Hanford's weapons chemical separation facilities—was refitted and put back into operation. One public reaction was the formation of the Spokane-based Hanford Education Action League (HEAL), whose work led to the 1986 disclosures of earlier off-site impacts from Hanford's plutonium production.

After more than a decade of growing public concern about nuclear activities—driven by the Three Mile Island accident, reactor siting battles, the search for potential national repository locations, and a new emphasis on weapons production—more Americans were engaged, with more negativity, regarding nuclear issues than ever before.[4] Then in April 1986, when the Chernobyl disaster affected millions of people, nuclear issues received even greater media and public attention.

Focus Shifts to the State of the Nuclear Weapons Complex

With the increasing media coverage and growing public concern, the attention of politicians, activists, and journalists turned to the nuclear weapons manufacturing complex with such questions as: Were these facilities being operated safely? What had been the environmental cost of four decades of "hidden" fissile materials production?

These questioners were further emboldened after 1986 with the rapid decline of the Soviet Union, as national secrecy and security concerns now appeared to be less important.

Congress—and particularly Senator John Glenn's oversight subcommittee of the Government Operations Committee—held extensive hearings. Experts from watchdog groups, such as the Natural Resources Defense Council, weighed in. Whistleblowers from various DOE sites, aided by the Government Accountability Project, provided hair-raising examples of poor safety practices.[5] A "midnight raid" by the FBI at Rocky Flats was perhaps the most dramatic event in this period. This eventually led to guilty pleas of violating environmental laws by contractor Rockwell's managers.[6]

As noted earlier, the states were beginning to assert "environmental" jurisdiction over nuclear sites, and the Department of Energy also began to work within the confines of the National Environmental Policy Act. Future decisions pertaining to nuclear facilities and waste management now required the preparation of assessments, subject to public review, regarding the extent to which policies and actions would "significantly affect the environment."[7]

At Hanford, for example, DOE determined that it needed to address the management of high level wastes stored in aging underground tanks and to prepare for the shipment of transuranic wastes to the Waste Isolation Pilot Plant under construction near Carlsbad, New Mexico. The resulting draft of the Hanford Defense Waste Environmental Impact Statement provided unprecedented focus

Nonprofit Watchdog Groups Identified in *America's Nuclear Wastelands*

National

Alliance for Nuclear Accountability (ANA)—1987; national/regional/local coalition of peace/environmental groups

Committee for a SANE Nuclear Policy (SANE)—1957; now combined in Peace Action (see below)

Energy Communities Alliance (ECA)—local governmental groups impacted by DOE policies and activities

Government Accountability Project (GAP)—1987; public interest group promoting government and corporate accountability

Natural Resources Defense Council (NRDC)—1970; environmental advocacy group

Nuclear Control Institute (NCI)—1981; advocacy center for preventing nuclear proliferation/terrorism

Nuclear Weapons Freeze Campaign (NWFC)—early 1980s; now combined in Peace Action (see below)

Peace Action—1987; combination of SANE and Nuclear Freeze

Physicians for Social Responsibility (PSR)—1961

Resources for the Future—1952; public policy, economics, law, and resources regarding critical environmental issues

Women Against Nuclear Armament

Regional

Fernald Residents for Environmental Safety and Health (FRESH)—Ohio

Hanford Education Action League (HEAL)—Washington

Hanford Watch—Portland, Oregon

Snake River Alliance—1979; Idaho

and information, and DOE convened a panel of prominent citizens to review it. Moreover, the State of Washington had in place both the technical staff and public involvement tools to review the draft, due to the state's ramping-up as a potential national repository host under the Nuclear Waste Policy Act.

As the review progressed, Hanford's future in plutonium production became increasingly problematic. The N-Reactor—the only reactor still producing plutonium—happened to be shut down for servicing in 1986 when the Chernobyl disaster occurred. Many critics pointed to the similarities in design between the two reactors. DOE disagreed, but appointed a scientific panel to review the N-Reactor's status. The panel concluded that a number of upgrades were necessary. By that time, however, the Soviet threat was fading fast, and the reactor would not be restarted.

In 1987, Hanford's PUREX plant also was stopped in mid-operation; one safety parameter (said to be 10 times more stringent than the applicable regulation) was exceeded. PUREX, too, would never be restarted.

Similar developments occurred elsewhere. Over the previous decades, federal uranium-processing and enrichment facilities had come to support not only nuclear weapons production, but also provided fuel to research reactors at domestic and foreign universities, and even for some commercial power generation. By the late 1980s, it was apparent that enriched uranium was abundant, that there was less commercial demand than expected, and that its continued use in research reactors constituted a significant weapons proliferation threat. Therefore, DOE closed aging and contaminated uranium-processing facilities at Fernald, Ohio, and Weldon Spring, Missouri, both of which had been encompassed by growing metropolitan areas.

DOE also concluded that the nation had much more plutonium on hand than was now required for weapons purposes. Therefore, the plutonium finishing activities at Rocky Flats were no longer needed. This complex of facilities had increasingly become the focus of anti-nuclear activity and local environmental concern as the Denver-Boulder metropolis engulfed it.

Past and Future: Jobs, Cleanup, and New Missions

The resulting extensive transitions generated deep conflicts in most of the communities located near defense nuclear sites. These municipalities provided the homes for a highly professional and relatively well-paid workforce, and a number of the scientists and engineers were intelligent and articulate defenders of things nuclear. Many felt that the cleanup did not present the same kind of challenges as their past employment; nor could it be regarded as carrying forward a nuclear future. In fact, cleanup appeared to admit past errors and to put a lot of these people on the defensive. However, there were a significant number of other people in these communities who thought that cleanup and waste management did indeed provide worthy challenges. The federal government soon would be investing several billion dollars annually into these activities, and scientific and engineering breakthroughs were sorely needed. The service sector in these communities, of course, could support the cleanup, just as it had the previous nuclear weapons production activities.

The conflict often took on the form of a debate over continuing nuclear missions: Did a site have a long-term nuclear future? For a number of the facilities engulfed by spreading urban areas, the answer obviously was a "No"; Weldon Spring, Fernald, and Mound at Miamisburg, Ohio, never were likely candidates for new missions. Rocky Flats, due to its large size, seemed less clear for a time. In all these cases, the surrounding local communities, backed by state governments, moved toward cleanup and the conversion of these locations to non-nuclear uses.

At Oak Ridge, Savannah River, Los Alamos, and the Idaho National Laboratory, local communities and state

Workers analyzing and characterizing waste drums during cleanup at the Weldon Spring Site, Missouri, April 26, 1996.

political leaders remained committed to supporting nuclear-based missions. As the Waste Isolation Pilot Plant in New Mexico began its successful operation, community leaders in Carlsbad eagerly sought additional nuclear-related activities. Not surprisingly, DOE tended to look to those areas that appeared to be relatively friendly territory as preferable sites for new missions.

The situations at the Nevada Test Site and Hanford were, and remain, quite complex. Each location deserves a more in-depth treatment than can be presented in this book due to space limitations, but following here are the basic issues.

After the changes of the 1980s, the State of Nevada has been in an uncomfortable and rather conflicting political position for years. State leaders supported the continuation of nuclear-related activities at the Nevada Test Site, which is a significant employer in the southern part of the state. In recent years, for example, the Nevada leadership has favored consolidating fissile materials at the test site; they maintain that the location is more secure than Los Alamos or other facilities. Nevada's government also has accepted the test site's continuing role in low level nuclear waste disposal in return for increased DOE funding to detect and control groundwater contamination there. But, on the other hand, Nevada's political leaders have almost unanimously and very vigorously opposed the locating of the deep national geologic repository at Yucca Mountain, situated on the southwest boundary of the Nevada Test Site.

Workers utilize a glove-bag during waste retrieval in the Fernald Closure, October 31, 2002.

At Hanford, the question of pursuing nuclear production missions continued to divide the adjacent Tri-Cities populace (Richland, Kennewick, and Pasco) and other municipalities in the surrounding region. In the late 1980s, many in the local community supported a restart of the N-

Reactor. When the decision came in 1989 to not further invest in the N-Reactor, the advocates of continuing production turned their attention to the PUREX facility. At a minimum, their argument ran, it should be restarted to clean out the production lines filled with dissolved nuclear fuel from when processing stopped in mid-operation in 1987. Furthermore, PUREX could be used to dissolve the spent fuel stored in the K-Reactor basins. DOE, however, opted instead to pursue an accelerated project to close up, decommission, and "safe-store" PUREX.

There was little debate, on the other hand, over the fact that the Pacific Northwest National Laboratory, operated by Battelle, would have a continuing presence at Hanford, conducting nuclear-related and other types of research. The PNNL laboratories were not production facilities and included a good deal of environmental and cleanup projects as well as defense and security activities.

Reversing the Weapons Race: A Hanford Case Study

By the mid 1990s, as the federal government determined that it had a surplus of plutonium for weapons manufacturing,[8] DOE began considering the approaches to adopt to reduce the stockpile down to what was needed for post-Cold War requirements. U.S. policy also was directed toward encouraging former states of the Soviet Union to follow suit.

DOE and other players began considering two related Hanford facilities as possible places for the disposal of the nation's surplus plutonium. One, the Fast Flux Test Facility (FFTF), a sodium-cooled nuclear reactor, had been built in the early 1970s to support the nation's research in fast breeder power reactors. The Fast Flux facility was designed to test various kinds of reactor fuel, as well other materials that might be used in reactor construction. As a research facility, the FFTF reactor was unusually flexible in the kinds of fuel that could be used, and was designed so that fuel and other "target" materials could be easily inserted or extracted. After President Carter's decision to shelve the fast breeder reactor program, the FFTF had operated only sporadically in specific materials-testing experiments.

The FFTF reactor's potential plutonium disposition role would be to burn mixed oxide (MOX) fuel—i.e., uranium fuel enriched with some plutonium. "Blending down" surplus plutonium into MOX fuel and burning it in civilian power reactors was one method of plutonium disposition accepted by the U.S. government, in part in order to draw the Russians into a similar program.[9]

Many Hanford scientists and engineers believed that the FFTF reactor previously never had the opportunity to fulfill its potential. Moreover, it was a civilian reactor, and never a part of Hanford's plutonium production mission. Its use to produce fuel for civilian purposes while destroying surplus plutonium would not, therefore, "put Hanford back into the weapons production business." Nor would it violate the strongly-held view of those concerned with international non-proliferation policies that strict separation be maintained between nuclear weapons work and civilian nuclear activities.

Supporters of this role for the FFTF reactor made considerable efforts to reach out to Hanford critics and the environmentally concerned political forces based in the more populous, western part of Washington state. Private investors put forward a number of proposals to take over the FFTF reactor, operating it to generate power using mixed oxide fuel, as well as to produce specialized radioactive isotopes needed in medicine and industry.[10]

Another Hanford facility, the Feed Materials Evaluation Facility (FMEF), had been built to support the FFTF reactor's original mission. It consists of a large, open storage structure containing in its center a number of "hot cells"—rooms that were shielded and equipped for remote disassembly and testing of highly radioactive materials, including fuel irradiated in the Fast Flux reactor. Built to the same "nuclear grade" specifications as a reactor facility, the FMEF is a massive concrete structure that can withstand earthquakes, tornados, and other significant calamities. The cessation of the breeder reactor program meant that the Feed Materials building had never contained radioactive materials. Its status as a "clean" facility meant that it could potentially be commissioned for another purpose without extensive regulatory investigation or cleanup expense.

DOE considered FMEF for a number of plutonium disposition missions. It might be used, stated a draft environmental impact statement, for disassembling metallic plutonium "pits," taken from surplus weapons, for blending MOX fuel, or preparing the plutonium to be mixed with liquid high level waste from Hanford or Savannah River. The combined plutonium and tank waste would then be vitrified and sent to the national geologic repository. A version of this disposal method also was accepted by the government as part of its program to dispose of surplus plutonium.[11]

As DOE considered its options for plutonium disposal in the mid-1990s, a number of groups in Washington state, including the League of Women Voters, Physicians for Social Responsibility, and the Tri-City Economic Development Council, worked together to hold forums on the issue near Hanford and in Seattle.[12] Speakers ranged from a senior scientist from the Nuclear Control Institute—a national nonprofit organization concerned about all forms of nuclear proliferation—to Russia's deputy minister for atomic energy.

The fundamental argument joined in these sessions was this: Should the plutonium be regarded as an asset or a dangerous waste? Those holding the former view, including the Russian minister, stressed that the material had been produced at great cost and represented a significant energy resource. Those opposed, including the Nuclear Control Institute representative, countered that converting the plutonium into mixed oxide fuel and converting reactors to burn it would also incur great costs. Further, this approach would sustain a "plutonium economy," in which the fissile material, potentially usable in weapons, would continue to be produced over the long run.

As a result of follow-up discussions at these forums, many organizations and individuals who had steadfastly opposed further nuclear missions at Hanford were now open to some role for plutonium disposition at the site. However, there was not much increased support for burning mixed oxide fuel in the Fast Flux Test Facility reactor. In the end, the pro-nuclear forces chose to back the use of the FFTF reactor to burn mixed oxide fuel, rather than support other potential plutonium disposition missions at Hanford. DOE, however, chose to locate all the disposition activities—pit disassembly, mixed oxide fuel fabrication, and plutonium disposal in high level waste—at the politically friendlier Savannah River site.

Demolition charges drop two reactor stacks at Hanford on June 30, 2002. This was one of the first activities performed as the reactors were placed in Interim Safe Storage (ISS).

Continuing Missions and Cleanup

One area of concern for state officials and others in regard to facilities without continuing missions has been this: Once DOE no longer has a long-term stake in a place, it will tend to divert resources elsewhere, and conduct only the absolutely minimum cleanup necessary in order for the department to walk away from a site.

Some observers, therefore, argue that securing ongoing missions will help cleanup efforts in the long run. On the other hand, others argue that the costs and obligations of soliciting continuing missions (and gaining the corresponding economic benefits) is too high; at the least, securing ongoing nuclear activities at sites may require political and regulatory compromises. Certainly, DOE has tended to propose the siting of new missions where politicians and regulators are perceived to be friendlier or "more reasonable."

Implicit in new missions—including for new activities and facilities related to cleanup efforts—is the awareness that mistakes of the past must not be repeated. Therefore, construction and operations need to be undertaken within existing environmental regulations, and with full consideration for the ultimate decommissioning and closure of various facilities.

Whether or not a site will have continuing nuclear missions, the cleanup must address two core questions that will be the focus of the next chapter:

• How clean is clean enough for future purposes?
• How can we be sure future generations are protected from harm?

Notes

1. USDOE, "Record of Decision: Management of Spent Nuclear Fuel from the K Basins at the Hanford Site," issued March 4, 1996.
2. A portion of U-235 is contained in enriched uranium and is much more fissile than naturally-occurring U-238.
3. "Disposal of Defense Waste in a Commercial Repository," memorandum from President Ronald Reagan to Energy Secretary John Herrington, April 30, 1985.
4. Eugene A. Rosa and William R. Freudenburg, "The Historical Development of Public Reactions to Nuclear Power: Implications for Nuclear Waste Policy," in Riley E. Dunlop, Michael E. Kraft, and Eugene A. Rosa, eds., *Public Reactions to Nuclear Waste: Citizens' Views of Repository Siting* (Durham, N.C.: Duke University Press, 1993).
5. For an excellent well-documented overview of this period, see Michele Stenehjem Gerber, *On the Home Front: The Cold War Legacy of the Hanford Nuclear Site* (Lincoln: University of Nebraska Press, 1992, 2002), pp. 4–9.
6. However, the lead FBI agent and grand jury members subsequently asserted that the environmental lapses were more severe than publicly revealed, and they sought to have the grand jury's report released. Neighboring property owners and former workers pursued civil claims, and a jury awarded substantial damages to the latter in August 2006. See Patricia Calhoun, "Flats, Busted: Lawsuits Against Rocky Flats, Like Plutonium, Last Forever," *Denver Westword*, March 9, 2006.
7. The early efforts by the Department of Energy to assess the scope of its environmental problems are described in *Complex Cleanup: The Environmental Legacy of Nuclear Weapons Production* (Washington, D.C.: Congressional Office of Technology Assessment, February 1991).
8. Strictly speaking, the Soviet-American agreements resulting from the Strategic Arms Limitation Treaty (SALT) talks in 1972 and 1979 limited nuclear weapons delivery systems, but not fissile materials. However, the reduction of the plutonium stockpile appeared prudent for many reasons, including reducing the threat of nuclear proliferation.
9. This approach is now being pursued, with MOX fuel scheduled to be used in Tennessee Valley Authority reactors.
10. The most serious proposals did depend on including a near-term defense mission—the production of tritium, which is used in nuclear weapons, but does not have a long

half-life. The concept was that a near-term purchase of tritium by DOE would cover costs until MOX fuel was available and commercial power could be marketed.

11. The appeal of mixing the plutonium with high level waste—or using a "can-in-can" method of encapsulating it in vitrified high level waste—is that the highly energetic radioactivity of the high level waste presented a significant barrier to retrieval and reuse of the plutonium for weapons purposes.

12. Washington Physicians for Social Responsibility and the League of Women Voters of Washington, "Plutonium Roundtable: Risks and Solutions—A Public Education Project on Policy Choices for Nuclear Weapons Disposal," October 5–7, 1995.

5

How Clean Is Clean Enough?

During the early 1990s, people in the northwest Denver metropolitan area fiercely debated the future of the Rocky Flats site. Located at the foot of the Front Range of the Rocky Mountains, between Denver and Boulder, Rocky Flats had become surrounded by urban development after the facility's creation in the early 1950s. Many who worked at Rocky Flats, as well as other residents in surrounding communities, feared the loss of high-paying

Workers at the 881 Hillside Hot Spot, Rocky Flats, September 27, 1994.

jobs, and deplored the possible elimination of a dense, sophisticated industrial complex should the site be closed.

Other people, however, were deeply concerned about the highly contaminated facilities and the potential for wind and water runoff to carry plutonium and other toxic materials into the increasingly urbanized landscape. They wanted Rocky Flats closed down and cleaned up—thoroughly. The latter group won out, for the most part, and Rocky Flats was officially cleaned up in the fall of 2005.

Why "for the most part"? First, because not all of the contaminated soil was removed from the site. After more fierce debate, regulators and DOE agreed that the completed soil cleanup would preclude any future "intensive" land uses; there would be no houses, farms, and schools at Rocky Flats. By act of Congress, Rocky Flats instead would become a wildlife refuge, and, within the site, the nearly 400-acre former industrial area would be restricted from human entry. Additionally, active long-term measures would be required to assure that runoff from the site did not enter streams supplying water to surrounding communities.

This outcome reflected a broad (but by no means universal) consensus that preserving nearly 10 square miles of open space in a sprawling urban area had, in itself, high value, and that removing "most" of the contamination and waste, and clearing away dangerous old facilities, would be the priority.

At the Mound site in Miamisburg, Ohio—a much smaller facility—the community coalesced around a cleanup plan that provided for making available some of the existing buildings and utilities for other industrial purposes. The cleanup standards were guided by an assumption that the industrial users would not be full-time residents, and, as working adults, they are less vulnerable to contamination hazards than children or older people. A higher standard of cleanup would have been needed at Mound if houses, schools, or farms were allowed there.

These cases illustrate that the answer to the question "How clean is clean enough?" varies from site to site, and reflects anticipated future uses at these locations.

If 50 years into the future one was to return to the top of Gable Mountain overlooking the Hanford Site, what might be seen?

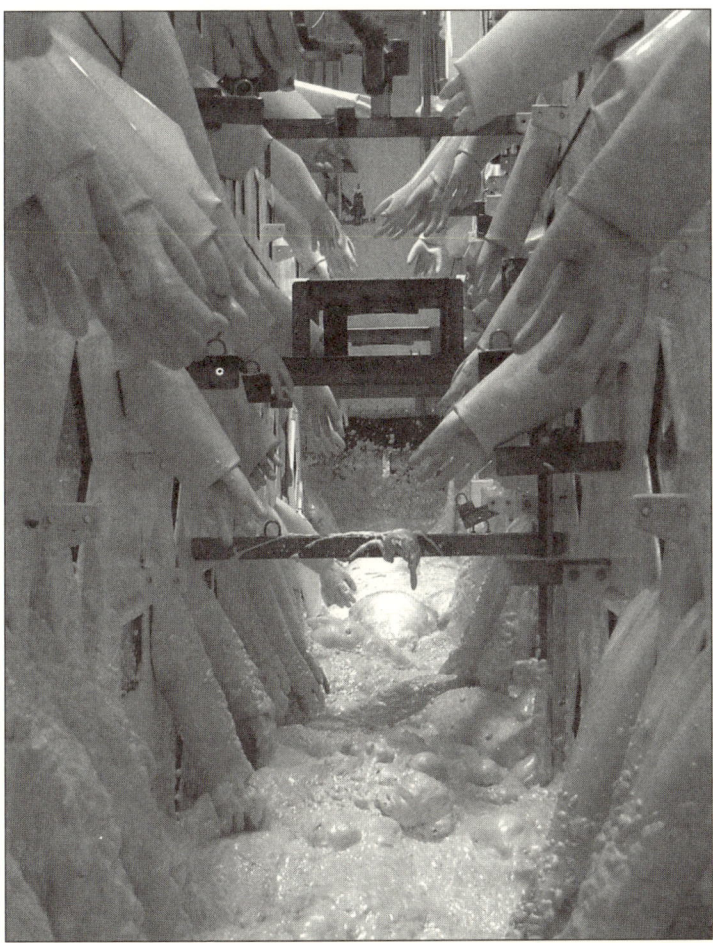

Foaming Building 371, glovebox 38, stabilizing contaminants from workers' gloves during the Rocky Flats cleanup, October 31, 2002.

Assuming that all of the old weapons production and waste treatment facilities had disappeared, would there be farms and houses as there had been in 1943? Or irrigated agriculture? Resorts, parks, and trails along the Columbia River? New types of industrial facilities just to the south in the former nuclear fuel reprocessing and waste management areas? Or, would the latter be covered over by monolithic earthen mounds, extending over three square miles and surrounded by high fences?

In the case of Hanford and other large sites with more varied waste and contamination profiles, including Oak Ridge, Savannah River, and the Idaho National Laboratory, a consensus that balances cleanup and future uses has not been reached—and it will be much harder to achieve where large regional resources, such as the Columbia River or the Snake River Plain Aquifer, are at stake. The continuing debates reflect deep divisions over values, and revolve around three basic issues:

- What should be the future use of the land and subsurface resources?
- How can people be protected from residual contamination?
- How long must we assure that they are protected?

At the most primary level, this is a debate ranging between those who want sites cleaned up to "greenfield" status so that potential future activities are totally unconstrained by the nuclear waste legacy, to those for whom "waste to wilderness"[1]—i.e., significant restricted future use—is the only sensible answer. Included in the latter scenario are anticipated future plans for wildlife habitat and limited recreational activities. Cleanup at this latter level does not allow for agricultural, residential, commercial, and industrial uses (i.e., activities that would expose people to residual risks, such as when digging, drilling, or irrigating).

Conflicting Values

Here are some of the key juxtaposed value conflicts that come into play:

Highest and best use?

- Cleanup should be so sufficiently effective that future uses of the land are not constrained by any residual contamination. The interaction of market forces and local planning should determine future activities, just as in any other land use situations.

- Or, cleanup should meet regulatory requirements that make it protective *at the boundaries* of a site, and control access to, or activities on, the site itself, in order to prevent uses that could expose people to the risk that still remains there.

Protection of future generations?

- Cleanup should protect human health and the environment for many generations into the future. Our children's children should not bear the consequences of our mid-20th century folly.
- Or, future generations should be required to set their own risk reduction priorities, and allocate resources accordingly. If at the present time we over-commit resources to eliminate the risk of nuclear contamination far into the future, we are using up resources and other means with which future generations might choose to attack risks.[2]

Restoration of resources?

- Cleanup should restore land and water resources to productive use, or, if that is impossible, compensate for the loss.
- Or, resources were consumed or contaminated in pursuit of national defense, which was an overriding public purpose at the time. Other material and human resources also were and are expended for national defense, thus complete restoration should not be expected as an end in itself.

Whose decisions anyway?

- Local communities, interest groups, Indian tribes, and state governments, all of which have a stake in the resources that are involved, must live with the results. Therefore, they should have a major say in defining likely future uses and the appropriate level of cleanup activity.
- Or, the federal government, as both the land owner and responsible party, must play the major role. This is because the government must balance the expenditure of tax monies among many competing demands across the nation.

Before we explore how these issues are addressed, it will be helpful to recall a couple of points from earlier chapters. First, there are a limited number of solutions in dealing with contaminated soil and groundwater: It can be scooped up and put somewhere else, or attempts can be made to contain it in place, with measures to keep it from spreading into the environment and populace. Second, there is a regulatory basis for defining a boundary between a contaminated site itself and the surrounding area; different considerations may apply on opposite sides of that boundary.

Completion of the Lip Area remediation at Rocky Flats, 2004.

For further clarity, if contaminants are present in the soil under or surrounding a storage tank, cleanup may mean scooping up soil and hauling it to an engineered landfill, where liners and caps prevent contaminants from escaping. Alternatively, grout or curtain walls and a cap may be used to contain contaminants in situ. If the latter approach is adopted, then continual monitoring will be required at the outside of the barrier, and there must be some reliable way to institute corrective action if contaminants begin to escape. Regulatory decisions will severely restrict surface uses above such a contaminated area, but may permit most normal surface uses outside of that particular facility's boundary.

Highest and Best Use

At the beginning of the 1990s, two developments discussed in earlier chapters propelled a flurry of future land use studies at nuclear defense sites. Policy adjustments after the end of the Cold War meant that many sites either would be assigned different missions or no missions at all, and the application of Comprehensive Environmental Response, Compensation and Liability Act (CERCLA) cleanup regulations required that attention be given to future land uses.

Regarding Hanford, the State of Washington, DOE, and EPA jointly commissioned a broadly representative Future Site Uses Working Group.[3] Parallel efforts were undertaken at Rocky Flats, Fernald, and Oak Ridge, and DOE developed guidelines for considering future land uses after cleanup for all of its sites.[4]

The Hanford effort had some notable features. Local governments, affected Indian tribes, and regional labor, business, agricultural, and environmental interests were represented in the Future Site Uses Working Group. The working group's report, a first attempt to relate decisions for cleanup and future land uses, contained these provisions:

- It divided Hanford into six areas, and indicated how different considerations would apply in each of these localities.
- Recognized that different timeframes would apply to surface and groundwater cleanup, as well as to cleanup in specific areas.
- Projected likely and desirable land uses for the different parts of the site, based on a broad array of public values.

The working group tempered the overarching goal of attaining "unrestricted use" with the recognition that this only might be achieved in the long run. The proposed future-use scenarios did not include residential and intensive agricultural development for any of the areas. These land use scenarios were subsequently refined in an officially adopted Hanford Comprehensive Land Use Plan.[5]

Based on its assessment of values, the Future Site Uses Working Group recommended that DOE and the regulators accelerate the cleanup of non-radiological contamination in Hanford's large buffer zones. The Arid Lands Ecology Reserve to the west toward Rattlesnake

The 300-Area at Hanford, looking west across the Columbia River with Rattlesnake Mountain on the horizon.

Mountain, and Wahluke Slope north of the Columbia River, indeed presented some low, very localized risks to public health and the environment. However, the group believed it was desirable that these localities be freed up for high-value habitat preservation and for recreational purposes. The working group also highlighted these same potential values along the Columbia River's relatively unspoiled shoreline, outside of specific areas where old reactors, laboratories, and fuel fabrication facilities had caused contamination.

DOE and its regulators acted on the working group's advice and accelerated the non-radiological cleanup of the two buffer areas. These, together with the river corridor outside of the contaminated areas, became the Hanford Reach National Monument,[6] which includes more than half of the Cold War-era Hanford Site.

The working group also recommended that waste and resid-ual contamination be concentrated in a central part of Hanford, encompassing the old industrial core where plutonium and uranium were extracted from spent nuclear fuel. As a result, DOE's contractors have moved more than 5 million tons of contaminated soil and debris from the old reactor and fuel fabrication areas along the Columbia River to the new Environmental Restoration Disposal Facility—a huge landfill located adjacent to the old industrial sector. Efforts to remove contaminates from sites along the Columbia generally were thorough enough to achieve the goal of allowing "unrestricted surface use." However, these efforts did not remove the subsurface risk that would be posed if digging, drilling, irrigation, or groundwater use occurred. Therefore, these activities, if undertaken, must be highly regulated in the future.

This is but one illustration of the interplay of a community's land use values with the cleanup activities that has occurred at several DOE sites. As in the case of Rocky Flats, however, some control (often called "institutional controls") over land uses are required for the long term to prevent contaminants from reaching people at levels currently deemed too risky.

The expression "long-term stewardship" emerged during the 1990s, denoting the need to maintain institutional controls and physical barriers for hundreds, or thousands, of years into the future. The meaning of "stewardship" in this context differs somewhat from the more common definition, wherein "stewardship" is

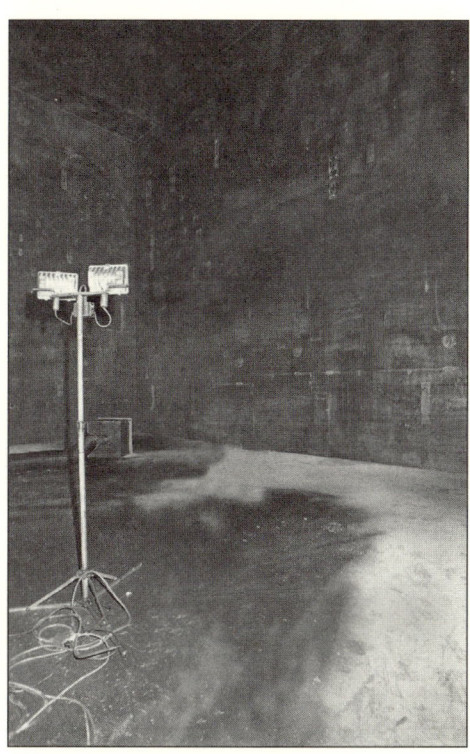

A view of 3563 "canyon" after the interior surfaces were sprayed with FireDam, a green water-based coating used to adhere the remaining contamination to the structure's surfaces; Rocky Flats Closure project, March 31, 2004.

understood to be the wise management of resources for sustained or increased yields in the future.

In order to better understand long-term stewardship at contaminated sites, we must further look at two of the basic issues presented at the start of this chapter—i.e., the restoration of an area's resources, and the protection of future generations.

Protection of Future Generations

Perhaps the best way to understand the value conflict that has arisen would be to envision a continuum. At one end, "cleanup" means making certain that future human generations need not worry about residual contamination left over from weapons production and other nuclear activities. All the residuals have been converted into non-harmful materials, or secured so completely that future generations are extremely unlikely to come in contact the them.

At the other end of the continuum, however, "cleanup" would mean securing contaminants with some combination of physical barriers and institutional controls, while waiting for either new and more efficient cleanup technologies, or new ways to re-use materials. Future generations would decide how best to further manage these materials and deal with the risk.

As a matter of policy, Congress has determined that the disposal of "high level" radioactive wastes and spent nuclear fuel plus "transuranic" wastes will be achieved in a manner closer to the continuum end that I first mentioned. To date, more than $8 billion has been invested in establishing the deep geologic repository in Nevada for spent fuel and high level waste,[7] and $3.9 billion for the development of New Mexico's Waste Isolation Pilot Plant, where transuranic contaminants are disposed.[8] The wastes at these sites will be isolated deep below the earth's surface, in extremely stable geologic strata. Within 100 to 300 years, these disposal sites will be sealed, and mankind will rely on the earth's secure geology to keep the wastes isolated. Thus, these repositories will be "passive"—i.e., nothing will have to be done to maintain them.

In the 1970s and 1980s, scientists investigated even more extreme isolation scenarios—for example, sending spent nuclear fuel and

Front loader covering low level waste packages at the Nevada Test Site, September 23, 1996.

high level waste into space, eventually to be consumed by the sun's energy. Another proposal suggested that wastes be deposited in the deep ocean floor, where over a great length of time it eventually would be carried into a subduction zone, where the earth's tectonic plates meet.

During the 1990s, Congress funded a modest research program exploring "transmutation," which are active processes that convert highly radioactive materials into either inert forms or into radioactive isotopes with much shorter half-lives and thus not requiring such long-term isolation. To date, this research has not yielded feasible processes that can deal with thousands of tons of nuclear waste.

In regard to "low level" waste, however, the opposite end of the continuum that I mentioned tends to pertain. Here are the essential characteristics of how the disposal of low level radioactive waste, mixed chemical/radioactive waste, and contaminated soil and debris are handled:

- Materials are disposed near the earth's surface, and thus might be potentially accessible to human contact and to dispersion by rain, wind, and fire.
- Isolation of these contaminants depends on waste containers, engineered liners, and/or caps that have relatively finite life expectancies.
- Over the last quarter century, the nation has adopted more stringent requirements for designing, monitoring, and maintaining facilities for these kinds of wastes, in part because many contaminants previously were inadequately "disposed" in landfills. Such wastes have had to be retrieved and re-disposed.
- Regulations tend to permit near-surface disposal of isotopes that have relatively short half-lives—i.e., isotopes that will be relatively harmless within, say, 300 years.

Keep in mind, however, the "Nuclear Waste Definitions" described in Chapter 3; "low level" Class C waste, for example, can be as hazardous as "high level" waste. Thus, some low level waste that might end up in a shallow-land disposal may resemble—or even be "hotter"—than materials that by law must go to deep geologic repositories. An anomaly in this regard can occur, for example, in a CERCLA-type retrieval cleanup, when the portion of a contaminant that has leaked out from a "high level" waste tank or transfer pipes is retrieved and sent to a landfill, whereas the same contaminant inside the tank is solidified and hauled to a repository.

Loss of Natural Resources

In the late 1950s and early 1960s, a weekly consumption of as little as one pound of whitefish taken from the Columbia River near Hanford may have exposed people to radiation doses exceeding the established standards (which, it must be pointed out, were becoming increasingly stringent over time). But the Atomic Energy Commission decided not to issue advisories regarding contaminated fish, such as those that have become familiar in later decades because of non-nuclear contamination in coastal harbors, such as Seattle's Duwamish Waterway or Tacoma's Commencement Bay.[9]

Today, the cessation of polluting operations at Hanford and radio-active decay over time means that Columbia River fish no longer contain potentially dangerous levels of radioactivity (though fish may retain unhealthy concentrations of other toxic materials from upstream agriculture and industrial mining and smelting).

The CERCLA law includes provisions for EPA to assess and compensate for natural resources damage due to the release of contaminants into the environment. Once the responsible parties have put cleanup measures in place, there must be some compensation for the natural resources that have been lost and cannot be restored.[10]

CERCLA's provisions are too complex to fully explore here, but two points relevant to this discussion can be presented:

- The law gives standing to other federal agencies, the states, and Indian tribes as trustees of natural resources.
- In the case of large, complex cleanups—such as those at nuclear weapons sites—it makes sense to begin the assessment of damages well before the final cleanup actions are in place.

Another example from Hanford illustrates how this is applied. Early in the 1990s, the Department of Energy (as both a liable party and trustee for natural resources under its control), the states of Washington and Oregon, the U.S. Fish and Wildlife Service, the U.S. Bureau of Land Management, and three affected Indian tribes formed the Hanford Natural Resource Trustee Council (NRTC). All parties have interests in the natural resources affected by Hanford's operations and its legacy. The Environmental Protection Agency's participation is ex officio.

The council's purpose is three-fold:[11]

- To help ensure that natural resources values are fully considered in decision-making related to the Hanford Site.
- To integrate, to the extent practicable, natural resources restoration into the cleanup actions and to minimize additional injuries to natural resources during cleanup.
- To encourage the development and implementation of site-wide natural resources planning by establishing mitigation, restoration, and management goals, and encompassing good stewardship practices.

The parties believed that being pro-active during cleanup operations could lessen the amount of damages that would need to be assessed at the end of cleanup. (Earlier, the Hanford Future Site Uses Working Group, too, had supported the principle of "do no more harm" in the cleanup.) The NRTC attempts to access the necessary technical perspectives to make this principle effective in practice. Thus early on, the council recognized the need to integrate natural resources information gathered from a myriad of studies and decisions affecting cleanup at hundreds of contaminated Hanford sites.

DOE, however, has found it difficult to balance its role as trustee with that of being the party liable for damages under the law. In fact, frustration with DOE's unwillingness to support trustee council assessments led the Yakama Indian Nation, joined by other tribes and the states of Washington and Oregon, to sue to force the department to begin funding comprehensive resource damage assessments.[12]

The extent of resource damage and loss is not yet known. Depending on one's perspective, it could be relatively minor—e.g., if compared to broad affects caused by agricultural and mining activities across the region over the past six decades—or the losses could be extremely large. The State of Washington has expressed particular concern about DOE's declaration that groundwater at the Hanford Site cannot be used for the foreseeable future. Washington law regards groundwater as a public resource to be allocated and regulated by the state.[13]

Long-term Maintenance of Protective Measures: Who Will Mind the Store?

As cleanup decisions are made at nuclear defense sites, the affected Indian tribes, state regulators, local governments, and a variety of interest groups often press for the goal of eventually achieving "unrestricted" use at these locations. In principle, long-term institutional and physical controls would not be needed if the cleanup was "complete."

There are several reasons for the desire to have a complete cleanup:

- Local communities must rely on government land-use controls to protect citizens and the environment from residual contamination, but these local institutional controls have sometimes proven unreliable. (A classic case of a failure of local governing overview—often cited—is the Love Canal site at Niagara Falls, New York. Closed in 1953 as a "chemical" landfill, the property was sold to the Niagara Falls Board of Education. Afterward, an elementary school was built on the site, as well as nearby residences, despite a recorded warning on the deed, indicating the presence of dangerous chemical wastes on this extremely contaminated property.[14]) In regard to the nuclear cleanup, to make matters more difficult, DOE's position has been that it cannot be bound by state and local laws to record, and make enforceable, such land use controls.[15]

- The repairing of structures, monitoring, maintenance, records management, and emergency response in the event of containment failures all cost money. DOE, as a federal agency, is not subject to financial assurance requirements under environmental law, which requires the establishment of trust funds or the posting of bonds. The department has taken the position that it cannot set aside funds to expend in the future.[16] Concerned parties fear, therefore, that DOE will not factor long-term costs into near-term cleanup decisions, thus biasing DOE's decision-making process toward accepting solutions requiring long-term stewardship.[17]

- American Indian tribes have treaty rights (long established under law) allowing access to natural resources at a number of their traditional localities situated outside of reservation boundaries. The tribes often assert that these rights have precedence in the remedial cleanup decisions focusing on long-term restrictions or prohibitions. The tribes also argue that any future spread of residual contamination will disproportionately affect them, because their diet, use of natural medicines, and spiritual or religious practices are closely tied to the natural environment.[18]

In short, these parties lack trust and confidence that DOE will act in a truly protective way and feel that the agency minimizes its near-term costs. On the other hand, DOE's view has been that cleaning to unrestricted use standards, even if it might prove to be technically feasible, does not focus efforts on reducing the highest current risks and it distorts priorities. A "Top-to-Bottom Review Team" chartered by Secretary of Energy Spencer Abraham delineated this position in a 2002 report:[19]

> Cleanup of the sites is often further complicated by a lack of realistic future land-use assumptions, and by scenarios that assume that highly contaminated areas will be subject to farming, drilling of wells, or residential use…Another major factor affecting USDOE cleanups is point of compliance for groundwater contamination. To the extent that the points of expected compliance with state and U.S. Environmental Protection Agency (EPA) standards are located near areas unlikely ever to be released for public use, unrealistic goals for cleanup are established.

To the untrusting, this statement meant that DOE believed it need not comply with certain environmental standards, and that it would determine future land uses. Suggesting that some land is "unlikely ever to be released for public use"—e.g., for agriculture, water sources, or homes—implied a degree of DOE certainty about long-term land use decisions that many observers believed was unjustified.

Earlier, a senior DOE cleanup official had responded to state and tribal concerns about the department relying on long-term controls to leave contamination in place:

> The level of cleanup is based on the expected future land use. We believe that if, after a site is remediated to levels appropriate for the specified land use, communities decide that they desire further cleanup to allow for less restrictive land uses, then the cost of such additional cleanup should not be borne by the Department of Energy.[20]

To skeptical local government officials, among others, this suggested that DOE would rely on its unilateral designation of future land uses to limit its liability, while thrusting the future costs of control and/or land use changes onto the local community. These are two of the reasons why long-term stewardship became a hot topic around defense nuclear sites in the mid 1990s. On one hand, there was a strong suspicion that DOE (and perhaps Congress and the Office of

Management and Budget) wanted to minimize near-term cleanup costs and shift long-term risks and costs to local communities. On the other hand, this DOE strategy required a reliance on assumptions about future land uses that many state, local, tribal, and other concerned stakeholders believed to be unrealistic.

If there is to be a heavy reliance on future land use controls and physical barriers, such as caps and underground trench liners, the next question is: Who is going to make sure that the necessary information is collected and maintained, needed inspections and repairs take place, legal restraints work, and these activities are sufficiently funded?

At sites such as Rocky Flats, Fernald, and Weldon Spring, DOE now has no further mission or reason to maintain an active presence. Furthermore, as noted earlier, a federal agency cannot commit to future spending. Thus, adequate future funding depends on annual Congressional appropriations and a political system trying to balance a myriad of crises and priorities. These questions are at the heart of long-term stewardship.

Shipment of transuranic waste leaving the Rocky Flats site, May 1, 2004.

In 2003, DOE created the Office of Legacy Management (LM), partly to deal with closed-down sites. Among the office's duties is to look after the maintenance of protective measures, legal restrictions, and records for those sites. This approach is based on DOE's earlier work done at closed uranium mining and milling sites. In its early days, however, the Office of Legacy Management office refused to press for more complete cleanup activities, in order to minimize its long-term costs.[21] Furthermore, the office did not have a new mechanism to assure long-term funding.

The Role of Indian Tribes

Tribal roles and perspectives are important in understanding the issues surrounding cleanup at many nuclear weapons sites. While I cannot speak authoritatively for the tribes, I believe I can indicate some of the important principles that they bring to the ongoing process of making cleanup decisions.

Just over 150 years ago, Euro-American settlers began to move into the far West, bringing very different ideas about land, property, and ownership, and often holding a condescending prejudice toward native peoples. In the face of inevitable cultural and land conflicts, the federal government negotiated treaties with the tribes. Today's tribal representatives generally emphasize three key facts about the treaties:

- They are "government to government" documents. The tribes, as sovereign nations, entered into these agreements with the United States.
- Under treaty terms, the tribes retained land for reservations (sometimes shrunk by later unilateral government action), but these documents also generally *reserved* the right of tribal members to continue their usual and accustomed subsistence activities in the larger traditional area that was ceded to the United States in treaty negotiations.
- Some tribes and bands were combined on reservations with little regard to their historical and cultural differences.

Affected American Indian Groups
Mentioned in *America's Nuclear Wastelands*

Confederated Tribes of the Umatilla Indian Reservation—
Oregon/Inland Northwest

Council of Energy Resource Tribes (CERT)

National Congress of American Indians (NCAI)

Navajo Nation—Southwest

New Mexico pueblos—Southwest

Nez Perce Tribe—Idaho/Inland Northwest

Seneca Nation—New York

Shoshone-Bannock Tribes—Idaho

Western Shoshone Tribe—Nevada/Southwest

Yakama Indian Nation—Washington/Inland Northwest

Two other widely-held tribal perspectives also are important to the discussion:

- Most Indian people feel a deep symbiotic relationship with the land. They depend on its natural bounty for sustenance, medicines, materials for everyday living, and for connection to a spiritual world. Consequently, tribal people often express a special responsibility in caring for the land, water, and biota that sustains them.
- They may move about from season to season, as different resources become available. However, all of the area they move in has been the tribal "home" to them for many centuries or millennia. Unlike

some Euro-American practices, they do not "use up" a place and then move on to another.

The federal government formally recognizes tribal participation regarding decisions about nuclear waste and cleanup in a number of ways. The Nuclear Waste Policy Act (1982)[22] grants specific rights to "affected Indian tribes." This "affected" status is based on a determination by the Secretary of the Interior that tribal rights might be impacted by repository site investigations. As a result, the Yakama Indian Nation (Washington), the Nez Perce Tribe (Idaho), and the Confederated Tribes of the Umatilla Indian Reservation (Oregon) began their long involvement with Hanford issues. Subsequently, the Western Shoshone Tribe was designated as "affected" by the development of the Yucca Mountain repository in Nevada. All of these tribes had exercised traditional usage rights in areas ceded by treaty and now under DOE's control.

Other tribes and pueblos affected by DOE facilities also participate in nuclear site cleanup under a Presidential Executive Order[23] and DOE's own American Indian Policy.[24] Among these are the Shoshone-Bannock Tribes in southeastern Idaho, with traditional lands in the Idaho National Laboratory, the northern New Mexico pueblos around Los Alamos and the Sandia National Laboratories, and the Seneca Nation, with traditional lands included in the West Valley, New York, site. These tribes, and a number of others, as well, including the Navajo Nation and New Mexico's southern pueblos, are involved regarding nuclear waste transportation routes across their reserves or ceded lands.

The federal government also has a trust responsibility toward Indian tribes, based on treaties and international law from the period of exploration and colonization.[25] This complex concept is subject to ongoing legal interpretations, but the crux is this: The United States is obligated to recognize tribal rights to natural resources and to self governance, free from state regulation.[26]

As a general rule, the tribes have relied on the federal trust responsibility as a primary basis to assert their interests in defense nuclear site cleanup. In regard to federal environmental laws, however, the tribes have no delegated authority over nuclear sites, unlike the states. In fact, the tribes are somewhat distrustful of state regulation, given

Debris sorting inside a waste management tent at Sandia National Laboratories, New Mexico, December 31, 2000.

historical conflicts over state efforts to regulate tribal affairs. The states and tribes, however, have found common ground regarding questions of natural resources damage and long-term stewardship at federal nuclear sites.

Long-term Stewardship Cannot Be Avoided

It should be apparent by now that a complete cleanup everywhere is not an option. A good deal of radioactive and chemical contamination will be with us, and our successors, for hundreds and thousands of years. Stewardship cannot be avoided. In a number of cases, hazardous materials will remain in long-term disposal at, or near, the surface of the ground, and thus potentially accessible to human intrusion or the affects of the elements. Land use controls, fences, physical barriers, and other means used to isolate contaminates can fail in the long run.

Disposal in the Effluent Treatment Project (ETP), Savannah River, February 11, 2004.

The longer-lived and usually more deadly radioactive contaminants will remain isolated, deep below the earth's surface in geologic repositories. Even these facilities, however, will require some level of monitoring and maintenance in the first hundred years or so. A good deal of thought also has gone into how to convey information about these repositories to the generations yet to come in future centuries.

Much of the discussion about nuclear waste and cleanup has a geographical focus. Will contamination stay in my backyard, or go to yours? Will the most intractable stewardship problems occur in Ohio, Washington, Idaho, or Utah? And who is at risk when moving contaminants from one backyard to another? The next chapter focuses on this geographic chess game.

Notes

1. Robert H. Nelson, *From Waste to Wilderness: Maintaining Biodiversity on Nuclear-Bomb-Building Sites* (Competitive Enterprise Institute, April 2001).
2. Milton Russell, *DOE Legacy Waste Cleanup and Stewardship: Beyond the Top-to-Bottom Review*, Joint Institute for Energy and Environment, Report No. 2002-06, August 2002.

The argument for carefully weighing present versus future costs and risks is clearly laid out in Part 3, "Principles for Managing DOE Wastes."

3. *The Future for Hanford: Uses and Cleanup*, Final Report of the Hanford Future Site Uses Working Group, Westinghouse Hanford Company, 1992.

4. Secretary of Energy Hazel O'Leary issued a "Land and Facility Use Policy" on December 21, 1994. The policy included the following statements: "We will integrate mission, economic, ecologic, social and cultural factors...Each comprehensive plan will consider the site's larger regional context and be developed with stakeholder participation."

5. USDOE, *Final Hanford Comprehensive Land Use Plan Environmental Impact Statement*, September 1999.

6. "Establishment of the Hanford Reach National Monument," Presidential Proclamation 7319 of June 9, 2000, *Federal Register*, June 13, 2000, pp. 37253–56.

7. Office of Civilian Radioactive Waste Management, *Annual Report to Congress, December 2004*, Accountability Report Appendix, p. 29.

8. The WIPP costs are stated for up through the 2006 fiscal year. The WIPP Information Center provided this figure to the author via e-mail, November 29, 2006.

9. Long-time residents I have talked to remember that Hanford employees somehow knew whitefish were contaminated, even though the AEC did not release a public advisory. For further discussion, see Michele Stenehjem Gerber, *On the Home Front: The Cold War Legacy of the Hanford Nuclear Site*, 2nd ed. (Lincoln: University of Nebraska Press: 2002), pp. 124–32.

10. For a full discussion of CERCLA provisions in this regard, see EPA's Natural Resource Damages Web page: www.epa.gov/superfund/programs/nrd.

11. Hanford Natural Resource Trustee Council, December 2005; see NRTC Web site: www.hanford.gov/public/boards/nrtc.

12. *Confederated Tribes and Bands of the Yakama Indian Nation v. United States Department of Energy*, CY-02-3105-LRS, Federal District Court for the Eastern District of Washington. In April 2007, DOE agreed in principle to begin funding natural resource damage assessment.

13. Revised Code of Washington, 90.44.035 and 90.44.040.

14. See, "Background on the Love Canal," Love Canal Collection, State University of New York at Buffalo, Library Archives: ublib.buffalo.edu/libraries.

15. Daniel S. Miller, "Looking a Gift Horse in the Mouth: Federal Agency Opposition to State Institutional Control Laws," *Environmental Law Reporter*, September 2002, pp. 1115–26.

16. "The federal budget process allows an annual debate about national priorities that results in funding appropriations for long-term stewardship and all other government functions. A number of commenters have expressed the concern that the annual budget process does not provide guaranteed funding for long-term stewardship activities...Because [other] alternatives...would require specific Congressional action in the form of legislation or specialized appropriations, the viability of these alternatives is depending upon Congress concluding that the annual budget process is inadequate for this purpose."—*Long-Term Stewardship Study: Volume 1—Report* (Washington, D.C.: USDOE, Office of Environmental Management, October 2001), pp. 98–102.

17. A concern about institutional controls is frequently found in the long-term stewardship literature. Probably most often cited is the discussion of constraints and limitations presented in Committee on the Remediation of Buried and Tank Waste, National Research Council, *Long-Term Institutional Management of U.S. Department of Energy Legacy Waste Sites* (Washington, D.C.: National Academy Press, 2000), pp. 52 ff.

18. These tribal concerns were outlined in a letter and memo to Secretary of Energy Spencer Abraham from the State and Tribal Government Working Group, sponsored by DOE's Office of Environmental Management, December 10, 2002. For a discussion of tribal lifestyles and cultural impacts, see Stuart Harris, "Cultural Legacies," a paper delivered at the Society of Risk Analysis annual meeting, December 7, 1998, Phoenix, Arizona, and included in Appendix B of Roy E. Gephart, *Hanford: A Conversation about Nuclear Waste and Cleanup* (Columbus: Battelle Press, 2003).

19. USDOE, Top-to-Bottom Review Team, "A Review of the Environmental Management Program," presented to the Assistant Secretary for Environmental Management, February 4, 2002, p. v–10.

20. Letter to State and Tribal Government Working Group, May 24, 1999.

21. Letter from Governor Gary Locke and Governor Bill Owens to Michael Owen, Director of the Office of Legacy Management, October 10, 2003. The letter and Owen's response are available at the National Governors Association Federal Facilities Task Force Web page: www.client-ross.com/cleanup-news/documents.htm.

22. Public Law 97-425, Section 117, provides for "consultation with states and affected Indian tribes," and Section 118 specifically with the "participation of Indian tribes."

23. *Consultation and Coordination with Indian Tribal Governments*, Executive Order 13084, May 14, 1998.

24. *U.S. Department of Energy American Indian and Alaska Native Tribal Government Policy*, January 2006, available at: www.ci.doe.gov/indianbk.pdf.

25. For a full discussion of the trust responsibilities, see *Long-Term Stewardship and the Federal Trust Responsibility: Incorporating the Duties Owed To and the Obligations of American Indian Tribes into a Long-Term Stewardship Plan, A Working Paper*, State and Tribal Government Working Group, December 2002.

26. *Johnson v. M'Intosh*, 8 Wheat, 543, 573-74 (1823).

6

The Geographic Chess Game

Governor Andrus' Opening Gambit

On October 20, 1988, Governor Cecil Andrus of Idaho sent the "tallest, meanest" state trooper available to stop a railroad car from entering the state. The boxcar carried transuranic waste from the Rocky Flats Plant in Colorado. Andrus' gambit opened an intricate policy-making game with the federal government—a game that ultimately engaged players from all across the United States. Similar games go on today.

Andrus' simple point: Stop shipping nuclear waste from other places for storage at the Idaho National Engineering Laboratory (INEL), located outside of Idaho Falls, when there was no commitment to move it out again.[1] Over the preceding years, thousands of barrels of transuranic wastes from Rocky Flats, as well as spent naval reactor fuel and other odds and ends, including damaged fuel from the ill-fated Three Mile Island power plant, had been deposited at INEL. Concerns had been raised about INEL because the sprawling facility sat above the Snake River Plain Aquifer, which was essential to much of Idaho's agricultural economy.

Andrus' action led, as one might expect, to litigation. In the meantime, the federal government did not defy Idaho's stand against the import of transuranic wastes. In 1990, however, the Department of Energy did attempt to bring spent fuel from an experimental civilian reactor in Colorado to INEL for storage. Andrus and the Shoshone-Bannock Tribes went to court to block those shipments.

In the end, Idaho won a federal court case that forced the federal government to complete an environmental impact statement on the INEL generally, and in regard to the disposal of spent naval fuel.[2] In the interim, the U.S. Navy stored its spent fuel at naval shipyards (Puget Sound, Mare Island, and Pearl Harbor). The Colorado Public Service Company meanwhile built a long-term storage facility for spent fuel at the Fort St. Vrain reactor.

By 1995, Andrus' successor, Governor Phil Batt, entered into an agreement with the U.S. Navy and the Department of Energy to have all of the related lawsuits resolved in a court-enforceable settlement. The final result was that DOE agreed to remove transuranic waste and spent nuclear fuel from the Idaho National Engineering Laboratory according to a specific schedule.[3]

The agreement meant that DOE either had to open the Waste Isolation Pilot Plant (WIPP) in New Mexico before 2000, or try to find yet another interim storage facility for transuranic waste. The mutual desire of DOE and Colorado officials to clean up and close the Rocky Flats facility also added pressure to complete the WIPP, which had been authorized by Congress more than a decade earlier. (Soon after Andrus had blocked further shipment of wastes to INEL, Colorado Governor Roy Romer had refused to extend the length of time permitted for storing mixed hazardous/radioactive transuranic wastes at Rocky Flats.)

Others also wanted to see the Waste Isolation Pilot Plant opened. Washington, Tennessee, South Carolina, and, to a lesser extent, California, Nevada, and Ohio, all had stored transuranic wastes needing long-term disposal. These states would not accept near-surface, on-site disposal at their in-state DOE facilities, since transuranic waste is harmful for many thousands of years.

New Mexico officials walked a delicate balance regarding the WIPP. Generally, the populace at Carlsbad and in southeastern New Mexico had long been boosters of the project. However, many environmentally aware and/or anti-nuclear individuals and groups in northern New Mexico, around the capital of Santa Fe, had strongly opposed WIPP.[4]

All these forces converged on Congress in the early 1990s. In October 1992, Congress passed the WIPP Land Withdrawal Act. The

Waste Isolation Pilot Plant near Carlsbad, New Mexico.

act balanced the desire to open the WIPP facility to receive transuranic wastes from the many states mentioned above with the desire by New Mexico to have adequate safeguards. The act included regulatory roles for the Environmental Protection Agency and for New Mexico, as well as provisions to support state and local transportation safety activities along transportation routes to WIPP.[5]

Northern New Mexico's Congressman, Bill Richardson, played a key role in brokering a legislative solution that recognized both the desire to open WIPP and the concerns of New Mexico and other states about federal over-bearance. Seven years later, Richardson, as Secretary of Energy, was on hand at WIPP to receive the first shipment of waste (from Los Alamos, in his former congressional district).

Federal Attempts to Site Waste Disposal Facilities

Faced with growing public concern and debate over waste issues, Congress made two efforts to structure the locating of waste disposal sites to overcome local and regional resistance. Both were subsequently modified in significant ways, and neither has worked out as intended.

(One only needs to recall the discussion of risk perception in Chapter 2 and the overlaying political geography to understand why seemingly rational schemes have not worked out well in practice.)

The two approaches, focusing on siting controversial facilities, cast the political dynamics of nuclear waste disposal into bold relief. They were not primarily concerned with the weapons complex cleanup; however, activists, advocates, and bureaucrats on all sides of defense waste disposal issues assume these experiences to be essential background. After briefly examining them, I will return to the interregional dynamics of weapons complex waste disposal.

Low Level Radioactive Waste Policy Act: The Limits of Interstate Cooperation

As the concern about radioactive waste disposal progressed in the 1970s, growing volumes of low level waste from hospitals, industry, and the increasing number of nuclear power facilities went to fewer and fewer shallow land waste disposal sites. As sites in New York, Kentucky, and Illinois closed, all the country's low level waste flowed to Nevada, South Carolina, and Washington.

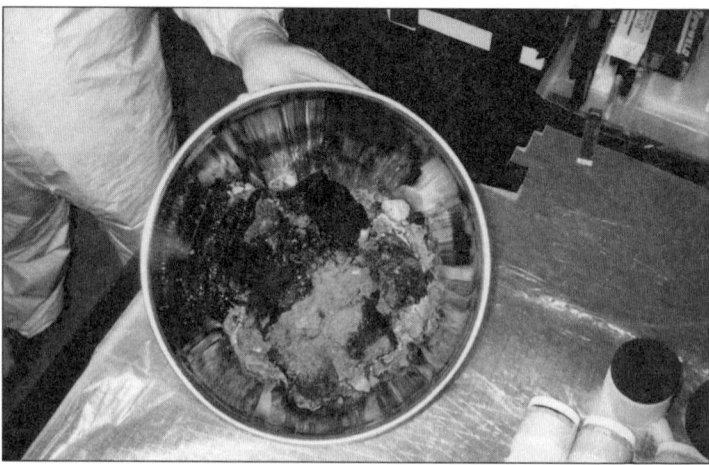

Typical waste material from the 183-H Basin at Hanford, May 31, 1994.

In Washington, even a very pro-nuclear governor, Dixie Lee Ray, felt compelled to crack down on unsafe shipments. Citizens there voted overwhelmingly in 1980 for the "Don't Waste Washington" initiative, prohibiting the import of out-of-state radioactive wastes. A federal court, however, soon overturned the initiative as interfering with interstate commerce. Meanwhile, the congressional delegations from the three states had shepherded through the Low Level Radioactive Waste Policy Act (1980). As mentioned briefly in an earlier chapter, the act permitted states to form interstate compacts that would select regional low level waste disposal sites and eventually exclude the import of wastes from outside the compact regions.

The assumption was that the states would join eight or so regional compacts, and within each compact the states would agree to a scientific and open process to select a desirable disposal location. A number of compacts were duly formed and authorized by Congress. Interestingly, a quarter-century later, only the Northwest compact—Washington, Oregon, Idaho, Montana, Wyoming, Alaska, and Hawaii—has an active disposal site. It is the same one, located within the boundaries of the federal Hanford Site, that already was operating before passage of the act.

The law was amended in 1985 to strengthen the compacts' authority to exclude out-of-region wastes after 1992, and to impose heavy surcharges on out-of-region waste before that time. Among the results of the surcharges, however, was a concerted effort by waste generators to minimize their volumes of waste. Consequently, the ensuing drop in waste reduced the economic viability of establishing additional waste disposal sites. As time went on, several states that originally expected to host regional facilities abandoned siting efforts when faced with rising costs and pressure from irate citizens. These included Pennsylvania, Michigan, North Carolina, Illinois, Nebraska, and California.

In addition to the falling volume of waste, three developments outside of the Low Level Act's scheme reduced the pressure for the compacts to develop new sites:

- The willingness of South Carolina, at least for a time, to operate outside the system and to accept waste from all over the country, using surcharge proceeds to help fund education.

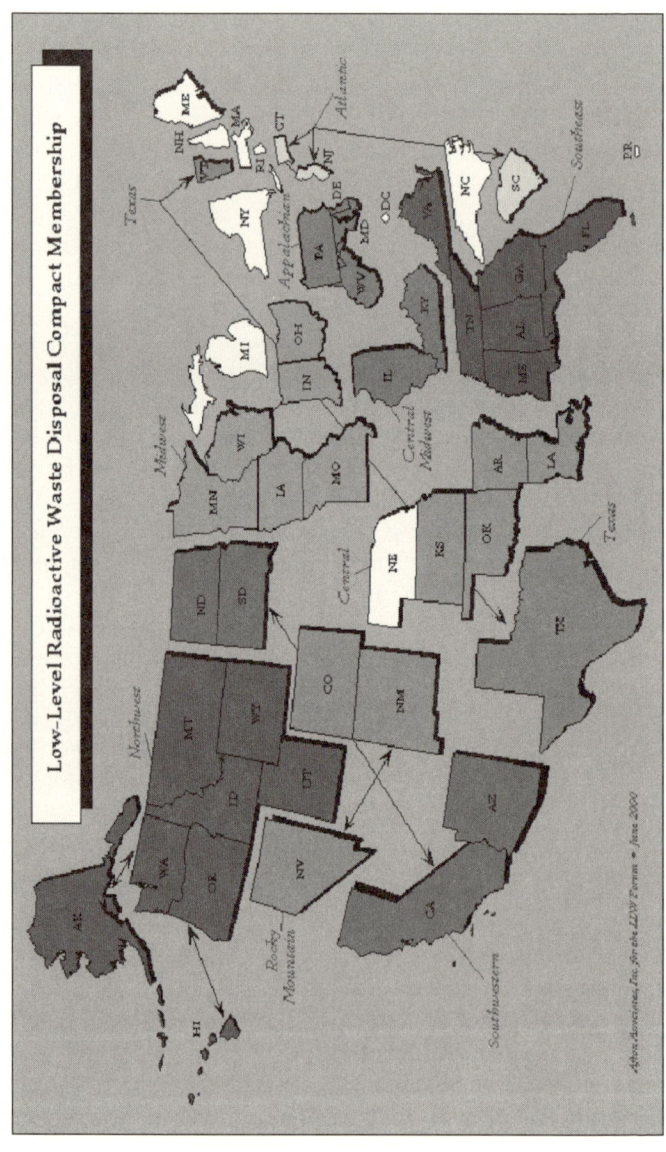

Figure 6.1. States in white are unaffiliated (Maine, Massachusetts, Michigan, Nebraska, New Hampshire, New York, North Carolina, and Rhode Island). *Courtesy of the Northwest Compact Executive Director, November 2006.*

- The development of a national disposal facility for bulk and low-activity waste in Utah, with the state's blessing.
- Continuing efforts to develop a site in Texas, either to support a "region" including California and/or upper New England, or to operate on a national basis.

Given these developments, the current compact situation looks rather odd (see Figure 6.1). Faced with the apparent malfunctioning of the system it created, and under pressure from utility and industry sources to fix the problem, Congress has nonetheless been reluctant to readdress the Low Level Radioactive Waste Policy Act.

This brief sketch illustrates three points that are significant for other areas of nuclear waste cleanup and disposal:

- Local and state political pressure—the "Not in my backyard" syndrome—has overcome both powerful interests (e.g., electric utilities) and scientifically based site evaluation and selection processes in many parts of the country.
- High costs and cumbersome public procedures often inspire new courses of action, such as waste volume reduction and developing new facilities outside the system.
- Congress must face intense pressure before it decides to revise such a controversial process as siting procedures for radioactive waste disposal facilities.

The Nuclear Waste Policy Act: Arrogance, Fear, and Federalism

As mentioned earlier, the federal Nuclear Waste Policy Act (NWPA) in 1982 created processes for siting and constructing national geologic repositories for spent nuclear fuel and high level nuclear waste. The act also included provisions to head off problems that Congress expected in the siting of such controversial facilities. This included step-by-step environmental reviews to narrow the number of potential sites to three; these then were to undergo extensive technical evaluation. Congress also provided for state and tribal oversight, and allowed for a state to veto the final selection of a repository site within its borders. Only a majority vote in both houses of Congress could override this veto.

In establishing this complicated process, Congress assumed that scientific credibility was crucial to the public's acceptance of repository siting decisions, and believed that involving state and tribal officials early in the process would enhance acceptance. Congress also wanted a regional balance by calling for two repositories—the first likely to be located in the West or the Gulf Coast; the other in the Upper Midwest or Appalachian states.[6]

The NWPA mandated that the site selection be completed by 1991, and the opening of the first repository would occur in 1998. In actuality, however, the president's official selection of a site (Yucca Mountain) did not occur until 2002—eleven years later than mandated. And even Congress itself had designated only one site (at the end of 1987). Early in 2006, Energy Secretary Samuel W. Bodman conceded that Yucca Mountain's opening date of 2012, proclaimed by his predecessor, was not reliable.[7]

Yucca Mountain exploratory studies tunnel, September 15, 1993.

So what went wrong? Why has this carefully crafted process failed to produce the intended result?

In 2001, a National Research Council committee concluded that "the biggest challenges to waste disposition are societal."[8] Regional politics and nuclear fear have been the major challenges. In my view, DOE failed to recognize and deal with these factors in its implementation of the NWPA.

In 1986, indeed, DOE selected three sites for detailed characterization: Deaf Smith County, Texas; Hanford, Washington; and Yucca Mountain, Nevada.[9] At the same time, DOE "suspended" (i.e., canceled) the search for a second repository site in the Upper Midwest and the East. It needs to be pointed out that of more than 100 nuclear power reactors operating in the United States at the time, most were in the eastern United States, and only four were in Texas, one in Washington, and none in Nevada. Westerners felt victimized.[10] Within a month of the decisions, the Western Governors Association unanimously condemned the cessation of a search for a second (Eastern) repository.

Representative Morris K. Udall of Arizona, the chairman of the Interior and Insular Affairs Committee at the time, summed up the perceptions of many:

> The Department of Energy had an unpopular and thankless task in selecting the sites, to be sure, but it had only to follow faithfully the process we laid down. The fact is…DOE blew it.[11]

As political circumstances would have it, the powerful Speaker of the House of Representatives came from a district near Hanford and the Majority Leader hailed from Texas, whereas Nevada had but weak representation with only two legislative seats in the 435-member House of Representatives. By the end of 1987, Congress had amended the Nuclear Waste Policy Act by directing DOE to characterize only the Nevada site at Yucca Mountain. In Nevada, the NWPA Amendments Act of 1987 was quickly dubbed the "screw Nevada bill."

One other provision of the 1987 amendments highlights how difficult it has been to choose a repository in the face of public fears. At the insistence of Representative Udall, the 1987 act included a provision for a Nuclear Waste Negotiator. The negotiator was given

Yucca Mountain exploratory tunnel, October 31, 1995.

broad authority to contact the states, tribes, and communities about potentially hosting a repository or interim storage facility, and to propose mitigation and compensation packages, including substantial monetary awards. Capable negotiators were selected in successive presidential appointments, but, nevertheless, no siting deals were ever proposed to Congress.

As a practical matter, proposals to provide mitigation and even substantial compensation, either to Nevada as repository host or to other states, has never shifted a critical mass of public opinion toward accepting national nuclear waste storage and disposal sites. The involuntary and dread aspects of the public's risk perception are quite resistant to such offers. In fact, state political leaders feared they would be vulnerable to charges of being "bought" if they even met with the Nuclear Waste Negotiator.

Two deeply-ingrained attitudes within DOE contributed to the problems of 1986 and 1987, and have continued to plague the focused effort to complete the Yucca Mountain Repository. (The State of Nevada has been adept at exposing these attitudes as part of its continuing opposition.) DOE, in a mindset typical of federal agencies generally, viewed its charge as finding a site that met standards set in part by itself, and in part by other agencies. It did not see its mission as necessarily finding the "best" site, or even a site that had especially good in-depth scientific credibility. Rather, once national political considerations had limited the selection to Yucca Mountain, DOE easily could be characterized as jiggering its science, its standards, or both to make the only "possible" site acceptable.[12]

Second, there was the arrogance of a relatively closed scientific and technical community. To a significant extent, both DOE personnel and the contractors working on repository siting and characterization had roots in the earlier defense nuclear complex. As noted, this community did not have a tradition of open scientific exchange with a broad range of disciplines. Here are some examples of the consequences:

- In DOE's view, the likelihood that contaminants buried in a repository would affect people on the surface was extremely remote, so the agency tended to dismiss the public's concerns in this regard as irrelevant. In part, DOE's confidence reflected its entrenched conviction that plutonium adheres to soil and would not travel in groundwater. Subsequent findings at the Nevada Test Site and Hanford suggest that this conviction was overly simplistic.[13]
- From the days of the Manhattan Project forward, many of those engaged in government nuclear projects believed an engineering solution to any arising problem could be found as work progressed, and cost was not a primary concern. Thus, as doubts arose about the capability of the geologic formation at Yucca Mountain to isolate contaminants for 10,000 years, DOE pursued a strategy of developing waste containers that would prevent leakage. The inherent difficulty in accurately predicting the performance of metal alloys over a span of 10,000 years is, perhaps, even more difficult than predicting geologic stability.[14]
- DOE's attitude has been that such factors as involuntary and dread risk are not scientifically valid, and therefore can be dismissed as

a chimera exploited by opportunistic politicians and anti-nuclear activists. This attitude has persisted even though a number of analyses in risk perception and communication, funded by DOE itself, have concluded that the public's risk perception is an important factor in how policy decisions are made.[15]

Congress attempted to deal with these attitudinal issues in its 1987 amendments. The amended law established a Nuclear Waste Technical Review Board, a multi-disciplinary group of scientists and social scientists appointed by the President. To the board's credit, it has provided opportunities for public involvement in its continuing review of the repository program. The board has produced a number of reports highlighting both the technical and social shortcomings of DOE's activities, though always from the point of view of improving the program, not eliminating it. The board is advisory only, however, and has no authority to direct DOE's conduct.[16]

The Politics of Defense Nuclear Waste Disposal

Waste Management Programmatic EIS

In 1989, the Natural Resources Defense Council and a host of other national and site-specific watchdog groups had sued Secretary of Energy James Watkins, claiming his department had no plan to deal with the waste and contamination at nuclear weapons production sites. In 1990, Watkins agreed to prepare a Programmatic Environmental Impact Statement (EIS) focusing on environmental restoration and waste management at DOE's complex of weapons-building facilities.[17]

All federal agencies are required to prepare detailed procedures to implement the National Environmental Policy Act, within the general guidance issued by the President's Council on Environmental Quality. Within this framework, the Department of Energy considered a programmatic EIS to be a broad-brush look at potential ways to configure a range of activities over time. Like all environmental impact statements, it provided opportunities for the public to review the information gathered about the environmental impacts and comment on the alternatives considered.

Watkins' successors in the Clinton administration carried through on most of this commitment. In 1994, DOE dropped the environmental restoration portion of the programmatic EIS, because decisions about the remediation of contaminated sites under the Comprehensive Environmental Response, Compensation and Liability Act (CERCLA) were made on a case-by-case, site-specific basis.[18] Thus, the document became the Waste Management Programmatic EIS.[19]

In this EIS, the Department of Energy explored many options for the treatment, storage, and disposal of high level radioactive waste, low level radioactive waste, transuranic waste, hazardous waste, and mixed low level and mixed transuranic waste. Because Congress had essentially decided where to dispose of high level waste (Yucca Mountain, in the amended Nuclear Waste Policy Act) and transuranic waste (New Mexico, in the WIPP Land Withdrawal Act), the EIS disposal options focused on low level and mixed low level wastes.[20]

Unlike the siting activities anticipated under the Low Level Radioactive Waste Policy Act (1980) and the Nuclear Waste Policy Act (1982), the Waste Management Programmatic EIS generally assumed that wastes would be disposed of at existing sites within the nuclear weapons production complex.[21] Thus, DOE proposed a range of configurations—from the disposal of wastes at nearly all of the sites (more than 50, including 17 major sites) where they had been or would be generated, to centralized disposal at a single site—Hanford or the Nevada Test Site.

The EIS considered several configurations for storage and treatment of wastes, particularly mixed radioactive and hazardous wastes, which (as noted in Chapter 3) are subject to state regulation. The document analyzed transportation impacts associated with the various configurations for treatment, storage, and disposal of the different wastes.

DOE held two major rounds of public meetings across the country. In the winter of 1990–1991, 23 public hearings focused on the proposed scope of the programmatic EIS; there also were 13 video-conference public hearings, involving 18 locations, on the Draft Programmatic EIS issued in 1995. In both instances, more than 1,200 individuals, organizations, tribal representatives, and government agencies provided comments. By any standard, this was a massive public involvement undertaking. In the same period, DOE also prepared and presented environmental impact statements related to its weapons production activities—to the disposal of surplus plutonium, to the WIPP facility, and to several site-specific decisions at Hanford, Oak Ridge, and other sites.

Attempts at dialogue

With the exception of the transportation of nuclear wastes, the public's input focused more on issues specific to local sites, rather than the big picture questions that the programmatic EIS was supposed to address. In part, these local reactions by stakeholders reflected the divisions that had arisen around each particular site regarding future missions, especially where other environmental impact statements relating to these potential missions were in process at the same time. In part, the comments often simply expressed opposition to the disposal of low level and mixed low level wastes at each local site.

Some skeptics would argue that this was an intentional divide-and-conquer approach that left DOE with a wide latitude to do what it already was predisposed to do. To be fair, however, DOE did support a number of efforts at broader, cross-site discussions.

In 1997, DOE's Richland Operations Office supported a pilot "national dialogue" project, designed by the League of Women Voters, the State of Washington, and interest groups ranging from the Physicians for Social Responsibility to the Tri-City Economic Development Council. The program included four workshops in Northwest cities, at which representatives from other areas hosting potential waste disposal facilities participated.[22] The State of Oregon also undertook a pilot process using smaller focus groups.

The following summer, the League of Women Voters Education Fund, supported by the Department of Energy, hosted two national dialogue workshops on waste treatment and disposal in San Diego and Chicago. Among other activities, participants contemplated the relative size and potential movements of waste inventories, represented by stacks of different colored Lego blocks on a large map of the United States.[23]

DOE also worked with representatives from the states by hosting waste-generating and potential-disposal-site discussions through the National Governors Association's Federal Facilities Task Force. This group originally had been created to deal with the movement of mixed wastes to various sites for treatment after the passage of the Federal Facilities Compliance Act (1992) clarified the states' authority over mixed waste treatment.

In all these cases, DOE attempted to stress both the national nature of the problem and the notion of reciprocity. For instance, for those concerned with Hanford, DOE officials repeatedly stressed that spent fuel, high level waste from tanks, and transuranic wastes, comprising by far the largest portion of Hanford's radioactive burden, would be shipped elsewhere for disposal. Even if low level and mixed low level waste from other sites were slated for disposal at Hanford, the net inventory of radioactive materials would be much lighter than at present.[24]

In the end, DOE decided to dispose of substantial quantities of low level waste at the major generating sites (INEL, Oak Ridge, Savannah River, Los Alamos, and Hanford), and to send the balance, as well as that portion of mixed low level waste not disposed of commercially, to the Nevada Test Site and Hanford. One of the major justifications for its decisions about low level waste disposal was that the amount of material to be transported would be considerably less than contemplated in several alternatives examined in the programmatic EIS. Officials claimed to have heard the public's concerns about transportation.[25]

Federalism and Conflict over Waste Disposal

The following brief accounts of recent conflict about waste management and disposal illustrate three key points:

- States are distrustful of DOE's efforts to define or classify wastes, because the definitions directly determine where wastes will be disposed of—at WIPP, at a national high level waste repository, or near-surface at nuclear materials production sites.
- States try to find common ground on jurisdictional issues even when their immediate cleanup interests may clash.

- The federal executive branch is consistent in its efforts to limit state authority in waste disposal matters regardless of party control. Congress and/or the courts are called on to mediate.

Hanford tank wastes to WIPP?

New Mexico's political leaders generally supported the opening and operation of the Waste Isolation Pilot Plant for transuranic wastes, with appropriate safeguards. However, they also have been vigilant to make sure that this waste disposal mission does not expand. WIPP, of course, is the only deep geologic repository currently operational in the United States

As the Yucca Mountain project in Nevada continues to languish, many New Mexicans fear increasing pressure to have WIPP accept

high level waste disposal as well. The lack of a high level waste repository also is a growing hindrance to the closure of some nuclear weapons production sites, de-commissioning or re-licensing of commercial nuclear power reactors, and public acceptance of new nuclear power plants.

Thus, federal proposals to expand WIPP's role to take spent fuel and high level waste would be politically explosive in New Mexico.

Hanford's Canister Storage Building (CSB) is critical to moving 2,300 tons of highly-radioactive spent nuclear fuel from water-filled basins near the Columbia River and into interim storage.

This is not only because of the public's concern about risk, but also because it would be viewed as reneging on long-standing commitments about WIPP's mission. Therefore, when DOE's Office of River Protection (ORP), which is responsible for managing and disposing of Hanford's tank wastes, suggested that some of that material might be defined as "transuranic" and sent to WIPP, Governor Bill Richardson of New Mexico took a "Hell, no" stance.[26]

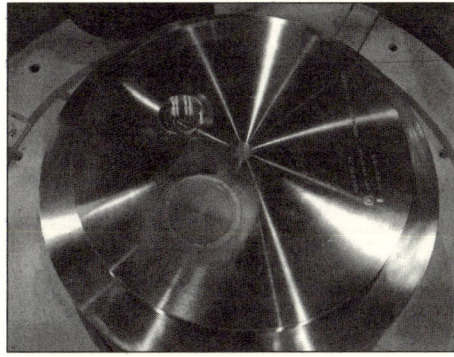

The first Shippingport Spent Fuel Canister (SSFC) is welded to N-stamp criteria in the Canister Storage Building (CSB) at Hanford. Welded SSFCs will remain in the CSB underground vaults until shipped to the national high level waste repository.

ORP asserted it could document that wastes in 7 to 20 of Hanford's tanks had not resulted directly from the processes by which federal regulations define high level waste.[27] From Richardson's perspective, however, all of Hanford's tank wastes had long been called "high level," and disposal of such waste at WIPP was unacceptable.[28]

Washington state officials were predisposed to support any efforts accelerating the retrieval and disposal of Hanford tank wastes, but did not side with DOE against New Mexico's resistance. One reason for this is Washington's continuing reluctance to accept DOE's assertion of unilaterally redefining wastes in order to change disposal paths. A second was a concern shared among several states, including New Mexico, about DOE's continuing attempt to restrict state authority over mixed transuranic waste.

On-site disposal of tank wastes?

In 2003, the Natural Resources Defense Council (NRDC) sued DOE for assuming the authority to redefine some high level waste as "incidental to reprocessing," and therefore exclude the material from the Nuclear Waste Policy Act's requirement for deep geologic disposal.

The issue arose because DOE had begun to use this assumed authority in deciding to leave some tank wastes in shallow-land disposal at the sites where they were generated. The NRDC was joined by the Snake River Alliance and the Shoshone-Bannock Tribes from Idaho, and the Yakama Indian Nation from Washington.

The states of Washington, Idaho, and South Carolina were most directly affected. In each, cleanup agreements assumed that some portion of tank waste would be disposed at the sites, but that the states would have a formal regulatory role. In South Carolina, the state had already agreed to the closure and filling of waste tanks with "heels"—i.e., hard residuals of waste left after pumping. In Washington, the major cleanup agreement always had anticipated that once the radioactive products of spent fuel reprocessing had been extracted to the maximum practical extent, the balance of tank wastes would be solidified and disposed on-site at Hanford.[29] Whether waste removal met the practical maximum extent threshold, however, was to be determined not by DOE, but by the Nuclear Regulatory Commission.

Idaho National Engineering Laboratory (INEL), December 1, 1993.

These states, joined by Oregon, entered the case as friends of the court. They agreed with the NRDC that DOE did not have authority to reclassify waste unilaterally, but they disagreed with the plaintiff's contention that the Nuclear Waste Policy Act required *all* tank waste to be sent to geologic disposal. An Idaho federal judge essentially ruled in agreement with the states.[30]

The federal government appealed the ruling to the 9th Circuit Court of Appeals. The states again filed a friend-of-the-court brief. This time they were joined by New York, which had concerns about the disposal of some wastes from West Valley, and by New Mexico, which as noted above wanted to avoid having tank waste reclassified as transuranic and therefore suitable for disposal at WIPP.

In the meantime, DOE officials told Congress that the "uncertainty" introduced by the decision in Idaho put accelerated cleanup decisions at the three major sites—and therefore funding to implement those decisions—at risk.[31] Congress subsequently attempted to find a legislative formula that allowed tank waste disposal decisions to go ahead. In the end, the formula adopted dealt only with South Carolina and Idaho, but not Washington. Meanwhile the 9th Circuit court overruled the Idaho decision, not on substance, but on the grounds that it was not "ripe" for judicial review.

State authority over the storage of transuranic mixed wastes

As noted earlier, states hosting DOE nuclear facilities have assumed that the Federal Facilities Compliance Act gives them authority to regulate mixed wastes (i.e., entailing hazardous chemical materials), including mixed transuranic wastes. DOE, however, has taken a different view. From its perspective, the storage and treatment of wastes destined for WIPP, including mixed transuranic wastes, are not subject to state regulation.[32]

Recall that the states' hazardous waste authority under the Resource Conservation and Recovery Act often includes limits on the time that hazardous chemical wastes can be held in storage before being treated and disposed. It was this authority that Colorado's Governor Roy Romer asserted to put pressure on DOE to open the WIPP facility in order to clean up Rocky Flats.

More recently, however, the issue turned to "remote-handled" transuranic waste—i.e., wastes that not only contain relatively low-energy, long-lived elements, but also highly energetic isotopes, so that it must be handled in a highly shielded way. In Tennessee, where Oak Ridge has a relatively large volume of remote-handled transuranic waste, DOE backed away from earlier commitments to remove the waste on a certain schedule, and asserted that the state could not require it do so.

It seemed DOE faced considerable delays in receiving the necessary regulatory approvals from New Mexico and the EPA for the Waste Isolation Pilot Plant to receive remote-handled waste. At the same time, DOE also wanted to accelerate cleanup at some smaller sites. One was a Battelle laboratory outside of Columbus, Ohio. In order to avoid constructing costly new storage facilities at the site, which was to be cleaned up before 2006, DOE decided to send the modest quantity of remote-handled transuranic waste, along with other wastes, to Hanford for storage until it could go to WIPP.[33]

In March 2003, the State of Washington challenged this decision in federal court, arguing, among other things, that the state had authority to regulate the storage of mixed transuranic waste. The federal government argued, on the contrary, that since the waste was determined by the Energy Secretary to be destined for WIPP, the state could not regulate its storage. Washington responded that it could not be certain the waste would go to WIPP, since final regulatory requirements for disposal of such waste at WIPP had not been established. Therefore, the waste might remain at Hanford indefinitely, outside the pale of hazardous waste regulation.[34] The district court eventually supported Washington's assertion of authority; however, the federal government appealed this ruling to the 9th Circuit.

"Rolling Chernobyl": Waste Transportation as a Political Issue

I will argue in the latter part of the next chapter that the safe transportation of nuclear waste is largely a success story. So why is it nonetheless a significant issue? The answer is primarily one of political geography. Most importantly, the use of highway and rail networks for hauling hazardous materials can be of immediate concern to more people than issues regarding the relatively isolated locations where waste treatment, storage, and disposal facilities are situated. That is why the discussion of potential transport routes through all the states and Congressional districts became a critical part of Nevada's campaign against the President's formal recommendation that Yucca Mountain become the nation's high level waste repository.

Indeed, most nuclear waste disposal sites are situated far from large metropolitan areas. And, as mentioned earlier, those smaller communities that have grown up around these sites are dependent on nuclear activities, and thus are more accepting of waste disposal.

However, people and organizations opposing the selection of a site for waste disposal (or other nuclear activities) often focus on the issue of transporting nuclear materials through large urban areas as a means to leverage broader political opposition. The perception of risk that is both dread and involuntary comes into play. The great majority of urban people who believe they might be affected by a nuclear waste transportation accident do not feel that their livelihood or community prosperity particularly benefits from nuclear waste-generating facilities and activities. The specter of an accident releasing invisible radioactivity over a large area—just as above-ground tests or reactor accidents did in the past—is scary indeed.

Thus, the potential to mobilize opposition is certainly there. When DOE examined alternatives for the return of spent nuclear fuel from overseas foreign research reactors, it considered a number of harbors for entry. Opposition was so intense around such ports as Tacoma, Washington, and Oakland, California, however, that DOE opted to receive the shipments at military terminals.[35]

The transport of nuclear wastes also became a political issue because state and local governments legally have some role. These jurisdictions both participate in the regulation of hazardous materials transportation and in the planning and preparation to respond to accidents. Therefore, local and state political leaders can be engaged—or choose to engage—in debates about nuclear waste transport and disposal.

The laws and precedents defining the limits of state and local, as against federal, authority in this arena are complex. Federal preemption, based on "interstate commerce," looms large, but state and local governments can and do enforce regulations adopted from, or not inconsistent with, federal rules. This is especially true with regard to roadway, as opposed to rail, transport.

In many cases, the federal rules become the means to mediate conflicts among the states and communities. Federal Department of Transportation (DOT) regulations concerning the highway-routing of radioactive materials arose from a long-running dispute involving the Brookhaven National Laboratory (Long Island), New York City, and New Jersey and Connecticut. Beginning in the 1970s, New York City sought to ban shipments of spent fuel from leaving a Brookhaven experimental reactor and passing through Queens and the Bronx on the way to the Idaho National Engineering Laboratory.[36] As this was the only land route, proposed alternatives involved barging the spent fuel to Connecticut or New Jersey, but officials in both of those states opposed such a plan.

The rule adopted by DOT prescribed basic standards for routing radioactive materials shipments, using interstate highways where possible, beltways around major cities when available, and minimizing the distance and time of transport. But it also provided a means for states to designate alternate routes. And it established a basis for DOT to decide whether state or local restrictions on transport were "not inconsistent with" federal preemption of transport regulation based on interstate commerce.

Generally, the federal government has accepted state and local regulations that are clearly tied to specific hazards—e.g., transport through long tunnels—or fees related to emergency preparedness. But the limits are narrowly drawn. In the late 1970s (before the federal regulations were in place), Governor Dixie Lee Ray of Washington ordered that all nuclear waste shipments coming into the state must enter at one of two ports of entry, both in eastern Washington. The more westerly of the two entry points was at Plymouth, located on the Columbia River just 35 miles south of Hanford. This meant that waste shipments from Oregon's Trojan commercial nuclear power plant—located some 200 miles to the west, on the south side of the

Columbia River—would have to be routed down a two-lane highway along the Columbia's south bank (the Oregon side) and through Portland, Oregon, on arterial streets until reaching the interstate freeway (I-84) in Oregon that would continue east to the Plymouth-area crossing of the Columbia River.

Oregon argued that federal rules required that the Trojan shipments should instead cross north of the Columbia near the Trojan plant and proceed through Washington state on the interstate highway (I-5) along the river's north bank, then proceed on a circumferential freeway around Portland to re-enter Oregon east of the city and gain access to Oregon's I-84 interstate route, to continue on to Plymouth. DOT agreed with Oregon and found Washington's requirement to be "inconsistent" and therefore preempted.[37]

In 1988, when Governor Andrus of Idaho asserted his authority regarding transportation to stop the boxcar shipment, it forced interstate and congressional action on Idaho's concern about accumulating wastes with no certain future disposal path. Had transportation been the real issue, Andrus probably would have lost the challenge regarding the consistency of his action with federal authority.[38] But, as is often the case, transportation was a means for a state official to weigh in on treatment, storage, and disposal issues.

Notes

1. Governor Andrus had failed to attain a commitment from DOE about the opening date for the Waste Isolation Pilot Plant in New Mexico. See Susan M. Stacy, *Proving the Principle*, DOE/ID 10799 (Idaho Falls: Idaho Operations Office, 2000), pp. 239–40.
2. USDOE, *Programmatic Spent Nuclear Fuel Management and Idaho National Engineering Laboratory Environmental Restoration and Waste Management Programs Final Environmental Impact Statement*, April 1995.
3. The issue of whether the agreement included the removal of *buried* transuranic wastes took the parties back to court. U.S. District Judge Edward Lodge ruled on May 25, 2006, that indeed it did. Former Idaho governors Andrus (a Democrat) and Batt (a Republican) provided testimony. Case No. CV 91-054-S-EJL.
4. Chuck McCutcheon, *Nuclear Reactions: The Politics of Opening a Radioactive Waste Disposal Site* (Albuquerque: University of New Mexico Press, 2002).
5. *WIPP Land Withdrawal Act* (Public Law 102-579).
6. The following discussion is based on the author's earlier paper, Max S. Power, "Politics and Science in Siting Battle," *Forum for Applied Research and Public Policy*, Fall 1989, pp. 19–25.

7. Mathew L. Wald, "Big Question Marks on Nuclear Waste Facility," *New York Times*, February 14, 2006.

8. Committee on Disposition of High-Level Radioactive Waste through Geological Isolation, *Disposition of High-Level Waste and Spent Nuclear Fuel* (Washington, D.C.: National Academy Press, 2001).

9. Earlier, the Nuclear Regulatory Commission and the affected states had sharply criticized DOE's approach to ranking the nine candidate sites for a first repository. DOE then agreed to use a more sophisticated method, called "multi-attribute utility analysis," to prepare its recommendation in narrowing down the sites; based on that method, a site in Mississippi ranked ahead of Texas, and Hanford ranked last. The later disregard of "scientific" comparisons dealt a serious blow to DOE's credibility in implementing the NWPA measures. Power, op. cit., p. 21.

10. On the first anniversary of the announcement of these decisions, DOE's Office of Civilian Radioactive Waste Management held a meeting in Las Vegas with state representatives and affected tribes from Texas, Washington, and Nevada. The state officials arranged for a cake, frosted in black, and a black-clad chanteuse, to "celebrate" the day.

11. "Is DOE's Waste Program in Trouble?" *Radwaste News*, Vol. 8, July 13, 1987.

12. This same criticism of federal agencies appears in a letter written by Robert R. Loux, Executive Director of Nevada's Agency for Nuclear Projects, and sent to the EPA when Loux was commenting on a proposed amendment to EPA's rule governing repository performance for 10,000 years. Loux argued that EPA was ruling out the consideration of "sound science." Numerous arguments along these lines may be seen at the Nevada agency's Web site: www.state.nv.us/nucwaste.

13. A.B. Kersting, et al., "Migration of Plutonium in Groundwater at the Nevada Test Site," *Nature*, Vol. 39, 1999, p. 756. I must thank Dirk Dunning of the Oregon Department of Energy's Nuclear Safety Division for explaining the chemical complexities that can lead to the solubility and mobility of plutonium in soil and groundwater.

14. Loux letter, op. cit.

15. For example, see Vincent Covello and Peter Sandman, "Risk Communication: Evolution and Revolution," in A. Wolhurst, ed., *Solutions to an Environment in Peril* (Baltimore: Johns Hopkins University Press, 2001).

16. See the U.S. Nuclear Waste Technical Review Board Web site: www.nwtrb.gov.

17. "Order by this Court on October 22, 1990," *Natural Resources Defense Council v. Watkins*, Civ. No. 89-1835 (SS) (1990 Stipulation and Order).

18. The EPA held that CERCLA procedures were "equivalent to" those required by the National Environmental Policy Act. Moreover, both EPA and the states believed that remedial action decisions at the sites were primarily theirs to make, not the Department of Energy's. Therefore they generally supported the decision to drop environmental restoration from the programmatic EIS.

19. USDOE, *Final Waste Management Programmatic Environmental Impact Statement for Managing Treatment, Storage, and Disposal of Radioactive and Hazardous Waste*, May 1997, DOE/EISW-0200-F.

20. Disposal of hazardous waste was discussed only briefly, with one clearly preferred option—continued disposal at commercial hazardous waste disposal facilities.

21. The exception was hazardous non-radioactive waste, which would be disposed, as it has been for years, at commercial hazardous waste disposal facilities. The program-

matic EIS also briefly discussed potential mixed waste disposal at commercial sites, but with little discussion of the specific sites.

22. League of Women Voters of Washington, *National Dialogue on DOE-Managed Nuclear Material and Waste: Pilot Field Workshops for Washington and Oregon*, Final Report with Attachments, Seattle, Washington, 1997.

23. League of Women Voters Education Fund, *Report to the Secretary of Energy: Inter-Site Discussions on Nuclear Material and Waste: A National Workshop*, December 1998.

24. Probably the clearest statement can be found in the Summary to the *Final Hanford Site Solid (Radioactive and Hazardous) Waste Program Environmental Impact Statement, Richland, Washington*, January 2004; Figure S.4 displayed current and potential waste and materials at, coming to, and leaving Hanford. The trouble with this claim was timing. Washington, for example, was being asked to commence receiving low level and mixed wastes from elsewhere, but it would be years or even decades before a repository to handle high level waste and spent fuel might be available.

25. "Identification of Preferred Alternatives for the Department of Energy's Waste Management Program: Low-Level Waste and Mixed Low-Level Waste Disposal Sites," *Federal Register*, December 10, 1999, pp. 68241–42.

26. "I want to make my position clear," Richardson stated. "I will not allow high-level waste in New Mexico—no matter what new name DOE comes up with to characterize it. This waste would be no different than it was before. Waste that has been defined as high-level for decades would suddenly become low-level on DOE's whim. WIPP was not designed or permitted to handle high-level waste, no matter what you call it."—Quoted in *The Raton Range*, October 31, 2003.

27. The technical argument is arcane. Suffice it to say that ORP argued that these wastes did not come directly from the first- or second-cycle reprocessing of spent nuclear fuel to extract uranium and plutonium. There was fairly broad technical agreement that the case for 7 of the small tanks, containing a relatively small inventory of radioactivity, was promising. There was much more doubt about the next 13 tanks on ORP's list. New Mexico attempted to avoid the technical morass by proposing a rule to limit WIPP to the reception of wastes specifically listed in a 1997 inventory included in DOE's last supplemental EIS on WIPP.

28. Governor Richardson's reaction also may have reflected the fact that the Department of Energy and the University of California—operators of the Los Alamos National Laboratory—were in court resisting New Mexico's efforts to regulate mixed waste management practices at Los Alamos.

29. South Carolina's position regarding tank residuals was based on its authority under the Clean Water Act, reflecting the fact that contaminants were in proximity to ground and surface waters. Washington based its decisions about tank closure on the Resources Conservation and Recovery Act, which essentially envisions the complete removal of contaminants from unsafe underground storage tanks.

30. *Natural Resources Defense Council v. Abraham*, District of Idaho, Case No. CV-01-413-S-BLW.

31. "The recent court decision in the Idaho District Court could significantly hinder our ability to implement the accelerated cleanup program." Testimony of Jesse Roberson, Assistant Secretary of Energy for Environmental Management, before the Investigations and Oversight Subcommittee of the House Energy and Commerce Committee, July 17, 2003.

32. Except insofar as New Mexico has authority to regulate the disposal of mixed waste at WIPP—a point addressed shortly in Chapter 6.
33. DOE apparently also considered sending the waste to Oak Ridge for storage. The State of Tennessee objected, however, by threatening not to authorize burn plans for a mixed waste incinerator at Oak Ridge, which was the only operating incinerator in DOE's complex and was essential for dealing with wastes from several sites.
34. *Washington v. Abraham*, Federal District Court for the Eastern District of Washington, Case No. CT-03-5018-AAM.
35. *Proposed Nuclear Weapons Nonproliferation Policy Concerning Foreign Research Reactor Spent Nuclear Fuel*, Final Environmental Impact Statement, February 1996. This EIS focused on a non-proliferation activity regarding foreign sources—i.e., the repatriation of fuel containing highly enriched uranium from which weapons-usable material might be extracted. The volume of foreign spent fuel was not large—about 19 metric tons, compared to about 2,500 metric tons already located at DOE sites. I attended a meeting that was called not by DOE, but by the Port of Tacoma Commission, to rally opposition.
36. The federal routing approach became known as "HM 164," named for the docket opened by DOT in August 1978. The final rule was adopted in January 1981. The *Federal Register* notice includes a fairly extensive summary of the controversies at the time. See *Federal Register*, Monday, January 19, 1981, p. 5298 ff.
37. Federal Highway Administration, "Preemption Determination Concerning State of Washington Port of Entry Restrictions and Their Effect on the Highway Routing of Radioactive Materials: Preemption Determination No. PD-3(F)," *Federal Register*, Vol. 58, No. 105, June 3, 1993, pp. 31580–87. The Trojan plant operated from 1976 to 1992. In 2005, the reactor vessel was barged to Hanford for burial. In the following year, the 499-foot cooling tower was demolished by explosives. The Trojan facility was the only nuclear power plant to operate in Oregon.
38. In fact, in the related case involving the transport of spent fuel from the Fort St. Vrain plant in Colorado (mentioned early in this chapter), Andrus and the Shoshone-Bannock Tribes lost on this point.

7

Cleanup Accomplishments

T he Department of Energy established the Office of Environmental Management (EM) at the end of the Cold War. By 2006, the program had expended about $90 billion.[1] In this chapter, I will summarize some of the resulting cleanup accomplishments. It is a bit misleading to say that this effort entailed the whole $90 billion.

Decontamination training in the Environmental Management program's Hazardous Materials Management and Emergency Response (HAMMER) at Hanford, May 24, 1995.

During the first decade and more, the Environmental Management program (and the budget) included a focus not just on cleanup and waste management activities, but also the maintenance of facilities and stockpiles of nuclear materials.[2]

DOE has, indeed, greatly reduced the risks of catastrophic accidents in regard to both the public and its workers. The management of critical nuclear materials has become more efficient, secure, and accountable. Cleanup is now complete at three relatively large and complex nuclear weapons production sites, as well as many smaller places. These sites are now, or soon will be, in long-term stewardship—the mode of continuing maintenance and surveillance.

After many delays and controversies, the Waste Isolation Pilot Plant (WIPP) in New Mexico is operating as planned; thus, DOE has been able to dispose of a significant portion of its backlog of stored transuranic wastes. DOE has made thousands of shipments to WIPP. It also has sent over a hundred unit trainloads of waste from Ohio to Utah, and transported significant quantities of other nuclear wastes and materials without any incidents leading to the harmful release of radioactivity.

For the most part, DOE's efforts to communicate with and to involve state, local, and tribal officials, concerned citizens, and workers have undergirded these achievements. Though the dialogue frequently has been messy and often seemed painfully slow, positive results are widely accepted. Ironically, as we shall see, these successes reduce the sense of urgency and the broad geographic scope of support necessary to meet significant unresolved problems.

The yet-to-be solved problems are many. These situations, if not addressed in the near term, will grow and pose more difficult challenges to future generations. For example, long-lived radioactive isotopes discharged into the soil in the past will continue (however slowly) to migrate—expanding the areas of contaminated groundwater. Tanks containing high level waste will continue to deteriorate. While the risk of catastrophic explosions may have been eliminated, leakage and the spread of contamination in the soil and groundwater becomes more likely as the tanks age. I shall return to some of these longer-term challenges in the next chapter.

Reduction of Significant Near-term Threats

As noted earlier, Congressional and public attention turned to the unsafe state of many nuclear weapons sites as the Cold War was winding down. Beginning in the mid-1980s, a number of serious near-term problems paraded across newspaper pages. At Hanford, a concern about possible chemical explosions in high level waste tanks was followed by the fear that a cooling-water failure could cause the igniting of spent fuel stored in old basins next to the Columbia River. Radioactive runoff in surface waters at Oak Ridge, Savannah River, and even the relatively arid Rocky Flats site generated calls for action.

Hanford workers utilize a radiological containment tent in the 244-AR vault, November 30, 2001. The vault had served as a waste storage and treatment facility for the plutonium-uranium recovery extraction plant (PUREX) in the late 1960s to early 1990s. Approximately 19,000 gallons of liquid waste remained in the storage tanks below.

Contaminated groundwater entered domestic wells in several places—adjacent to the Fernald Plant, near Cincinnati; at the Paducah Gaseous Diffusion Plant, in Kentucky; and later near the Brookhaven National Laboratory on Long Island. Only of little less immediate alarm, there was a concern that transuranic wastes buried at the Idaho National Engineering Laboratory would reach the Snake River Plain Aquifer, the mainstay of southern Idaho's agriculture.

By the mid-1990s, new threats appeared in the headlines. The media presented photos of a mound of rusting drums containing uranium wastes sitting on the ground and contaminating surface waters at the Paducah site in Kentucky. A chemical explosion occurred in the Plutonium Finishing Plant at Hanford. Meanwhile, the physical deterioration of facilities from the Manhattan Project and Cold War eras posed increasing risks to workers, over and above the normal concerns of exposure to radioactive materials or toxic chemicals. In April 1992, a Hanford worker inspecting the roof of an aging reactor building died when it gave way.

Congress became sufficiently concerned about safety at defense nuclear facilities that it created an independent oversight board, the Defense Nuclear Facilities Safety Board (1988). The board's members, appointed by the President, have investigative powers, staff resources, and the authority to issue formal recommendations to the Secretary of Energy.[3] The secretary has the power to reject the board's recommendations, but rarely has done so. The board reports annually to Congress regarding its recommendations and DOE's progress in regard to compliance.

The Defense Nuclear Facilities Safety Board (DNFSB) has since made a number of recommendations to reduce near-term risks. The recommendations have focused primarily on *how* nuclear-related work is done, whether for cleanup or restart of production facilities. The cleanup program has eliminated many near-term potential threats at Hanford, including tank explosions, accidents with unstable plutonium solutions, and the rupture of spent nuclear fuel basins near the Columbia River. The removal of drum piles and scrap metal at Paducah and Oak Ridge has decreased runoff contamination into surface and ground water.

The response to the near-term risk of tank explosions at Hanford is an example of just how complex these safety issues can be. As mentioned in Chapter 3, I was sent to Richland in the fall of 1989 to investigate after the media reported that ferrocyanide—combustible at temperatures above, say, 500 degrees Fahrenheit—had been used to precipitate cesium and strontium in a number of Hanford's older single-shell tanks. What I learned in my hurried review of the situation was this: The removal of cesium and strontium had reduced the generation of heat in the radioactive materials in these tanks; thus, ignition of the ferrocyanide now was unlikely. On the other hand, the thermo-couples—devices by which contractors measured temperatures—had been inoperable in several of the tanks for months or years.

At nearly the same time, gas generated by chemical decomposition in Hanford's storage tanks also captured public and regulator attention. One tank in particular, the "burping" 101-SY, frequently vented built-up pockets of potentially inflammable hydrogen gas.

U.S. Governmental Oversight Groups Established and Chartered by Congress

Defense Nuclear Facilities Safety Board (DNFSB)— 1988; Executive Branch advisory panel charged with oversight of past and present defense production.

Government Accountability Office (GAO)—since 1921.

National Research Council (NRC)—since 1916; National Academy of Sciences and other scientific and technical policy advice.

Nuclear Regulatory Commission (NRC)—1974; authority to license repository and related facilities and some oversight of defense high level waste.

Nuclear Waste Technical Review Board (NWTRB)— 1987; independent scientific and technical oversight of the repository program under the Nuclear Waste Policy Act.

President's Council on Environmental Quality (CEQ)—1969.

High heat and the presence of organic chemicals in other tanks added to the overall concern.

Keep in mind, too, that at the end of the Cold War, Americans learned more about incidents at similar facilities in the Soviet Union. New information was available about the effects of a 1957 waste tank explosion at Mayak, near Kyshtym, in the southern Ural Mountains. Highly radioactive waste dried out and ignited spontaneously, contaminating an area of about 30 by 200 miles, and causing the evacuation of up to 270,000 people.

Faced with these concerns, Congress created the Wyden Watch List (named for then-Representative Ron Wyden of Oregon) in 1990. The law required DOE to list all those tanks suspected of having potential problems and to take measures to minimize the risk, or demonstrate that the chance of catastrophic risk was extremely low.[4] Beginning in January 1991, 60 of Hanford's 177 high level waste tanks were listed. In just over 10 years, DOE's response and efforts would meet the requirements of the law in removing all of these tanks from the list, which was "closed" in August 2001.[5]

Workers pump out liquids from one of Hanford's older single-shell tanks (SSTs) on October 31, 2002. The pumping program began in 1998.

During that decade, DOE spent tens of millions of dollars to develop new technologies and equipment to sample tank wastes, reduce gas generation, and remove wastes. This included non-sparking tools, mixer pumps to prevent the build-up of gas bubbles in waste, and video cameras that could operate in a highly radioactive environment.

Since 2001, DOE has completed the task of removing free liquids from all of Hanford's single-shell tanks (SSTs), and has begun retrieving the peanut-butter-like sludge and crystallized salt-cake from those tanks. The possibility of a catastrophic event in a Hanford tank spewing radioactivity into the atmosphere is now extremely minimal.

So also is the threat of Hanford's spent fuel basins contaminating the Columbia River or that fuel will be exposed to the air and ignite. All 2,100 metric tons of spent fuel has been removed from the basins, dried, packaged, and stored in a specially-designed facility away from the river. Contractors are still at work removing water and sludge from the basins.

In 1987, plutonium processing at Hanford's Plutonium Finishing Plant had been

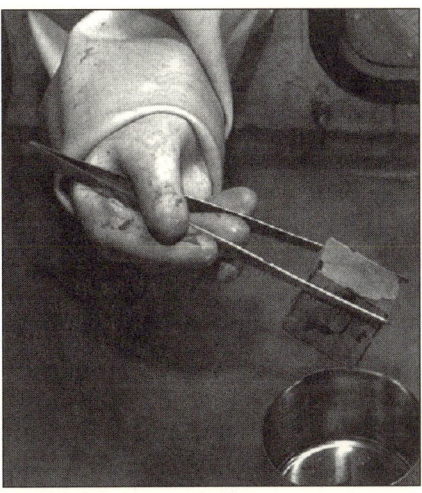

A Plutonium Finishing Plant worker handles a polycube inside a glovebox, March 31, 2003, in the process of stabilizing one of the riskiest forms of plutonium-bearing materials at Hanford. Polycubes are small cubes of polystyrene impregnated with pure plutonium oxide.

stopped abruptly when a safety limit (one-tenth of a regulatory limit) was exceeded. DOE, however, never restarted the plant because, with the winding down of the Cold War, it was determined that the United States had a surplus of weapons-grade plutonium. However, chemical solutions and other unstable materials containing plutonium, intended for short-term use, were left in place throughout the plant. These materials did not pose a major threat to surrounding communities, but their unstable state in an aging facility posed substantial and increasing risks to workers.

By the end of 2004, these plutonium solutions and materials had been stabilized, safely packaged, and stored in a facility subject to international inspection. As with the tanks and the removal of spent fuel from the old storage basins, this task took several years, hundreds of millions of dollars, and required significant technological advances.

This last effort benefited somewhat from techniques developed in cleaning up the plutonium-shaping facilities at Rocky Flats. But only somewhat. As noted in Chapter 1, facilities to produce nuclear weapons materials—and later to explore other applications of nuclear power—were developed at different times and locations in an extremely compartmentalized fashion. Activities at each major facility changed over time, not necessarily in sync with those at other locations.[6] Thus, stabilization and cleanup technologies have not been easily transferable.

The storage of high level wastes, perhaps, provides the clearest example. In the 1940s, when the initial chemical separation process of retrieving plutonium began at Hanford, stainless steel was unavailable for the site due to wartime shortages. Consequently, the tanks receiving highly acidic liquid wastes were constructed of carbon steel, which is susceptible to corrosion by acids. As a counter measure, large amounts of base materials—sulfates and sulfites primarily—were introduced to neutralize the wastes. This use of carbon steel tanks and basic solutions also continued into the Cold War.

At Savannah River, on the other hand, waste tanks were constructed from stainless steel, so that less neutralization was required. As a result, Savannah River's high level waste is more concentrated and acidic, but it also has precipitated far less salt cake and sludge than the Hanford tanks. At the Idaho National Laboratory, which chemically dissolved a smaller quantity of irradiated fuel, primarily from naval propulsion reactors, liquid wastes were dried into a granular material, called "calcine," rather like laundry detergent.

Therefore, the retrieval and treatment of high level waste stored in these various tanks has posed different technical challenges for each of the sites.

Consolidation of Materials

The initial compartmentalized and semi-autonomous field structure of the Department of Energy (and its pre-1977 predecessor agencies) had led to a dispersion of various nuclear and non-nuclear materials at many sites. The nuclear materials of greatest interest—uranium, plutonium, and spent nuclear fuel—raise legitimate health and safety concerns. For the most part, these materials also require a high degree of security due to their potential use in weapons.

The situation has been exacerbated because these materials were not only managed by different site offices, but also because similar materials might "belong" to different programs within the department's organization—e.g., Nuclear Energy, Naval Nuclear Reactors, National Nuclear Security Administration, or Environmental Management. In the latter half of the 1990s, the department undertook to identify and better manage these materials.[7] The next step—to consolidate like materials at a single location—is underway.

There are two reasons why this is important. First, the administrative and facility costs of managing smaller quantities of key materials at several sites are high. Second, the accountability for these materials, especially highly enriched uranium, plutonium, and spent fuel containing quantities of these elements, is an essential part of arms control and non-proliferation policies.

For example, the secure storage of surplus plutonium at Hanford costs nearly $80 million per year. In addition, because the material is stored in part of the Plutonium Finishing Plant complex, workers coming into the area to clean out and decommission buildings must be subjected to extra, time-consuming checks and clearances. Record keeping and the provisions for international inspections of the stored plutonium also impose additional costs.

Since the funds for Hanford's infrastructure, as well as security, come from the Office of Environmental Management budget, the continued storage of plutonium there decreases the pace of cleanup. Ultimately, plutonium will be disposed of in facilities at Savannah River. The consolidating of materials and the centralizing of security and international accountability at Savannah River makes sense.

DOE also has moved to consolidate highly enriched uranium stocks, as well as other uranium isotopes, at Oak Ridge. Thus, uranium stored at Hanford that had been intended for N-Reactor fuel was transferred to Tennessee.

Spent nuclear fuel from the department's various research reactors and naval vessels, as well as highly enriched uranium fuel from domestic and foreign research reactors, also has been scattered at several sites. The fuel comes in two basic types, aluminum clad and zirconium clad, each requiring different storage conditions. The fuel is now being consolidated at two sites—the aluminum clad at Savannah River, and the zirconium clad at the Idaho National Laboratory.

Completion of Cleanup at Major Sites

According to the Environmental Management program's Corporate Performance Measures in late 2005, DOE would complete cleanup at 78 of 114 sites by 2006, and expected to finish 12 more in the following three years.[8] Many of the 78 places were relatively small uranium extraction or storage sites or laboratories. The scale of these achievements had grown substantially by 2005 with the completion of cleanup work at Weldon Spring and Rocky Flats. Fernald was expected to follow in 2006. These latter three sites had extensive facilities and a complex set of contamination problems resulting from past operations, accidents, and inadequate waste disposal practices. All were located near growing metropolitan areas; all had significant surface or groundwater contamination issues.

DOE also has made significant progress in addressing long-term surface and groundwater contamination problems at Hanford, Oak Ridge, Los Alamos, and Savannah River. This has been accomplished by:

- Removing poorly-disposed waste from areas where it contaminated runoff—e.g., old, low level waste disposal trenches on a hillside at Oak Ridge.
- Dams to collect potentially contaminated water for treatment before reaching streams—e.g., at Rocky Flats, Oak Ridge, and elsewhere.
- Barriers to arrest groundwater movement and bind-up contaminants—e.g., underground chemically-treated dikes to extract chromium headed toward the Columbia River at Hanford.
- Pump-and-treat systems to extract contaminants from groundwater—e.g., the removal of uranium from the Great Miami Aquifer below Fernald.

Construction of the Dynamic Underground Stripping (DUS) system at Savannah River, June 30, 2003. This was the second deployment of DUS at the site targeting an area of over 10 million cubic feet, with the expectation of removing up to 1 million pounds of solvents in the former M-Area Settling Basin.

The Waste Isolation Pilot Plant in Operation

Completion of cleanup at Rocky Flats would have been impossible without the opening and effective operation of the Waste Isolation Pilot Plant near Carlsbad, New Mexico. Nor would DOE have been able to meet the deadlines for waste removal imposed by the Idaho Settlement Agreement (see Chapter 6). From 1988 to 1992, Governor Andrus and Congressman Richardson had rightly grasped the importance of WIPP.

In all, 2,045 shipments of transuranic waste went from Rocky Flats to WIPP.[9] More than 9,100 cubic meters (1,500+ truck shipments) also left Idaho for WIPP by the end of 2005, so that the Idaho facility could remain open to receiving spent fuel from naval ships.[10] Moreover, 250+ shipments of transuranic waste sent from Hanford made it possible to accelerate the retrieval of deteriorating waste drums from shallow trenches at Hanford.

Site-wide drill at Rocky Flats, October 14, 1998.

It appeared that WIPP soon would be permitted to receive remote-handled transuranic waste, after the completion of extended negotiations among DOE, the State of New Mexico, and interested others. Opening WIPP up to accepting remote-handled waste would help DOE meet cleanup commitments at Oak Ridge and Hanford, and to complete removal of wastes from laboratories in Ohio.

My view is that four factors contribute substantially to WIPP's success:

- *A geologically robust site.* The WIPP facility was developed in a geologic formation that was thoroughly studied. DOE and contractor evaluations underwent significant independent reviews throughout the permitting process in developing WIPP. The independent Environmental Evaluation Group, funded by DOE but appointed by New Mexico authorities, provided a continuous and thoroughly technical critique.

- *Close cooperation between WIPP and its regulators, the New Mexico Environment Department and the U.S. Environmental Protection Agency.* This has been neither easy nor always cordial, but the three parties have all been committed to having the facility operate safely within the framework of hazardous waste law.
- *Point-of-origin audits.* WIPP, with the participation of EPA and New Mexico regulators, has conducted frequent and thorough audits at the sites sending wastes to WIPP. At times, these audits have identified deficiencies that frustrate politically-sensitive commitments to ship waste from generator sites. However, this approach minimizes conflicts with regulators and the disruption of disposal activities at WIPP.
- *Ownership of shipments.* Once waste is packaged according to requirements, WIPP controls shipment to its facility. This provides at least three advantages. First, WIPP knows when to expect shipments and what they contain; thus, receipt and disposal operations are efficiently organized. Second, WIPP will know if anything has occurred during transport that would affect the acceptability of waste containers that had met requirements at the point of origin. Third, as I shall discuss in the next section, WIPP officials have long understood that transportation safety is key to public, state, local, and tribal acceptance of their operations.

Safe Transportation of Nuclear Wastes[11]

As of mid-2006, DOE had made 4,766 truck shipments—a total of more than 5.2 million miles of highway travel—of transuranic waste from eight sites to WIPP. These shipments came from as far away as Washington and South Carolina. There has been at least one vehicular collision involving a shipment, but no releases of radioactivity that were above, or even near, regulatory limits. Perhaps as importantly, relatively few persons or organizations have actively opposed these shipments. Also, after the first few shipments, there has been limited media attention. State and local officials were generally confident that such shipments, and thousands more to come, could be safely accomplished.

Several factors account for the safety and acceptability of WIPP shipments, and many of the same factors apply to other nuclear waste shipments as well. At present, the WIPP transportation process is viewed, at least by most state and local officials, as the best model. DOE and the Western Governors Association cooperatively developed specific WIPP transportation protocols during the 1990s. These protocols were tested in the mid-1990s, before WIPP's opening, when a Colorado irradiator sent capsules of cesium, taken earlier from Hanford's tanks, back to Washington state.

Here are the major factors that enhance the safety of WIPP transuranic shipments, both from radiological and traffic perspectives:

- *Containment.* This is the primary line of defense against the release of radioactive material in the event of an accident. The waste-carrying containers are tested and certified by the Nuclear Regulatory Commission to meet stringent standards. Containers, or casks, must withstand 1475° Fahrenheit of heat and fire for 30 minutes and a drop of 30 feet to an unyielding surface without rupturing or releasing the contents. The closure mechanisms for the casks and the devices that secure them to the trailers are designed to work together. The container type most often used for contact-handled transuranic waste shipment is the TRUPACT-II.
- *Driver training and certification.* The WIPP organization contracts with specific carriers to transport waste. Contractors select drivers based on their excellent safety records, who are then extensively trained. A single moving violation by a driver is grounds for dismissal.
- *Vehicle inspections.* The appropriate state officers inspect trucks carrying WIPP-bound waste at the point of departure, using inspection protocols agreed upon by an organization of all states, known as the Commercial Vehicle Safety Alliance. Standards for removing a vehicle from service under these protocols are stringent. States along the transportation corridors may conduct additional inspections, using the same standards.
- *Designated routes.* Shipments to WIPP follow designated routes. These generally comply with DOT regulations; however, state, tribal, and local governments have been involved in selecting the

At the Nevada Test Site, a shipment of low level waste contained in metal boxes awaits unloading and disposal at the Area-3 Radioactive Waste Management Site, June 14, 2005.

routes, and are consulted if any changes occur. Protocols agreed upon between the states' regional organizations and DOE also specify how to proceed with deviations from designated routes due to weather conditions, highway construction activity, and roadway-impacting accidents.

- *Real-time tracking.* Each shipment is tracked by satellite (TRANSCOM), both by WIPP and state and local officials along a route. In addition, drivers must call in regularly to confirm their positions.
- *Emergency preparedness.* DOE has provided resources and training for state, tribal, and local emergency responses along WIPP shipping routes. Training is frequently repeated (especially important in areas served by volunteer fire departments), and includes hospital and medical personnel.

- *Bad weather protocols.* The states and DOE have agreed on procedures to monitor weather and road conditions so that WIPP shipments can avoid particularly hazardous road conditions. Shipments do not depart DOE facilities if severe weather is likely to be encountered; trucks also are required to park if meteorological conditions unexpectedly deteriorate.
- *Public information.* WIPP officials, in cooperation with representatives from state, tribal, and local agencies, have actively informed the public about WIPP's shipping program. These people visit communities along transportation routes, often accompanied by a truck hauling TRUPACT-II containers. Thus, concerned citizens have an opportunity to question not only WIPP representatives, but their own state, tribal, or local emergency and law enforcement personnel about shipment safety. While the times of specific shipments are not announced publicly, anyone can easily become informed about the routes and the projected number of shipments on those routes.

The transportation of "low level" radioactive waste, on the other hand, occurs without these specific protocols. The Nuclear Regulatory Commission and the Department of Transportation, in this regard, have prescribed their own container requirements, which vary depending on the type and quantity of radioactive materials to be shipped. (However, recall that some "low level" waste may contain highly radioactive materials.) The Department of Transportation sets permitting, vehicle, driver safety, and placarding[12] requirements for interstate carriers of these and other hazardous materials. With the exception of small containers of radioactive chemicals used in medical diagnosis and treatment, the regulations prohibit the shipment of radioactive liquids. These nuclear waste shipments contain only dry, often solid, materials.

A number of states require a permit for the transportation of radioactive materials or wastes, and may inspect all or most low level waste shipments, as well as shipments of transuranic waste or spent fuel.[13]

Since the beginning of the nuclear age, there have been thousands of shipments of spent nuclear fuel and other highly radioactive materials—including the steel racks used for holding nuclear fuel inside reactors—over America's transportation networks. And, according to the Nuclear Regulatory Commission, no radiation releases harmful to the environment or public has occurred during shipping for over 30 years.[14]

The Necessity of Continued Vigilance and Engagement

The accomplishments recounted in this chapter should go some way toward reducing primal fears associated with the nuclear weapons cleanup legacy. Yes; catastrophic accidents that can spread radioactivity in the environment are much less likely now than two decades ago. Yes; many old, dangerous facilities and unsafe disposal areas have been cleaned up and waste removed. Yes; there is better management of fissile materials. Yes; nuclear waste transportation is relatively safe compared to the shipment of other hazardous materials.

But this record of improved safety and progress will not be self-sustaining unless interested citizens, the media, Congress, and state, tribal, and local officials continue to be concerned and focused. An irony of these successes is that some people have become less involved—the number of states and Congressional districts with difficult nuclear waste issues has decreased. Thus the forces that produced these successes are waning. Moreover, the generation of people who have tackled issues in the nuclear weapons complex cleanup is graying rapidly.

In the next chapter, I shall outline several complex, expensive, and long-running cleanup challenges. These are extremely difficult technically, but failure to address them now will increase the environmental and public health burden on future generations. Moreover, if the nation invests in new nuclear power facilities in the near future, it must do so in a way that avoids the mistakes of the past.

Notes

1. It is surprisingly difficult to arrive at an accurate estimate. Over time, DOE has changed both its budget categories and the use of current vs. constant dollars. This estimate simply assumes about $6 billion annually for 15 years.
2. In 2002, the Top to Bottom Review estimated 7% of the annual budget went to activities not related to cleanup. In 1999, a DOE official indicated to me that the amount spent on the maintenance of facilities and stockpiles, not related to cleanup, was on the order of 30%.
3. Title 42, *Defense Authorization Act for 1989*, 42 USC § 2286. For a description of the Defense Nuclear Facilities Safety Board and its recommendations through the years, see: www.dnfsb.gov. It must be kept in mind that when Congress divided the Atomic Energy Commission's functions in the Energy Reorganization Act (1974), the law excluded defense nuclear facilities from regulation by the Nuclear Regulatory Commission. Thus, DOE remained self-regulating with regard to nuclear weapons activities.
4. Section 3137 of the *National Defense Authorization Act for Fiscal Year 1991*, Public Law 101-510.
5. E.S. Aromi, et al., "Hanford Site River Protection Project High-Level Waste Safe Storage and Retrieval," paper delivered at Waste Management '02, Tucson, Arizona, February 24–28, 2002.
6. For a good summary of the different processes applied at different sites over time, see USDOE, Office of Environmental Management, *Linking Legacies: Connecting the Cold War Nuclear Weapons Production Processes to Their Environmental Consequences* (Washington, D.C., January 1997), Appendix B.
7. Ibid.
8. "Corporate Performance Measures at the Complex Level: EM Program," Rev. 15, December 2005. For these purposes, DOE defines completion as meeting all the requirements of the cleanup agreements with state and federal regulators. Stewardship activities and/or ongoing program activities may continue at a site eliminated from the Office of Environmental Management's list.
9. WIPP Web site, "Shipment Figures" table, updated June 26, 2006. For more current numbers, see: www.wipp.energy.gov/shipments.htm.
10. Per the terms of the 1995 settlement agreement, discussed in Chapter 6.
11. The following discussion applies primarily to the transportation of "non-defense" nuclear wastes and materials. The "safe, secure transport" of fissile and weapons components materials has its own set of provisions.
12. Federal regulations require uniform placards on trucks or rail cars carrying hazardous materials. These placards enable emergency response personnel to identify the specific hazards to be dealt with in case of an accident.
13. Generally speaking, states have more authority and latitude to impose fees, inspections, and regulations regarding highway shipments than they do to the rail and marine transport of radioactive materials. Two primary factors account for this: Railroads are private corporations (as opposed to the public ownership of highways), and the federal government has a long-established preemption of railroad and marine

regulation. Some states, however, do assess fees to support inspections and potential emergency responses related to rail shipments of nuclear waste. DOE generally has not challenged these fees so long as they meet a test of "reasonableness."

14. "Transportation of Spent Nuclear Fuel: What We Regulate": www.nrc.gov/waste/spent-fuel-transp.html.

Highly contaminated "canyon" at a Hanford plutonium separation facility, November 2, 2004.

8

Cleanup Challenges for Today and Tomorrow

The cleanup tasks remaining at nuclear weapons production sites are challenging indeed. DOE, with cooperation from regulators and stakeholders, has made substantial headway in reducing risks, razing or securing aging facilities, and shrinking the footprints of contamination on the land. But some of the most technically baffling and expensive tasks remain. These challenges raise important public policy questions. Successful resolution will require continued public engagement and a level of trust among concerned parties and the federal government.

Perhaps a mixture of metaphors will help indicate why critical challenges still bedevil cleanup. It has been said that DOE's program went after the "low-hanging fruit" first. While not entirely true, this conveys an important point. Moreover, in using another metaphor, turning the "ship's rudder" of fissile materials production toward a different course at the end of the Cold War has been as slow and difficult as turning any massive aircraft carrier or super tanker. The inertia was—and is—a force to be reckoned with.[1]

In short, it took some years to remold the entrenched attitudes, contracting mechanisms, and business procedures within DOE in order to focus on the cleanup. For a considerable time, the first response to cleanup challenges was to do things as they always had been done—e.g., a proposed way to dispose of spent fuel at Hanford was to restart the PUREX plant and reprocess the fuel; the way to deal with plutonium solutions in the Plutonium Finishing Plant (PFP) was to restart the facility; and the way to process irradiated fuel stored at Savannah River was to restart separation at F-Canyon.

Hanford Reactors and Major Facilities

	Start-up	Shutdown	Current Status
B-Reactor	Sept. 1944	Feb. 1968	Potential museum
D-Reactor	Dec. 1944	June 1967	Interim Safe Storage (ISS)/"cocooned"
F-Reactor	Feb. 1945	June 1965	ISS/"cocooned"
H-Reactor	Oct. 1949	April 1965	ISS/"cocooned"
DR-Reactor	Oct. 1950	Dec. 1964	In ISS process
C-Reactor	Nov. 1952	April 1969	In ISS process
KW-Reactor	Jan. 1955	Feb. 1970	Decommissioned/ ISS by 2011
KE-Reactor	April 1955	Jan. 1971	Decommissioned/ ISS by 2011
N-Reactor	Dec. 1963	Jan. 1987	Decommissioned/ ISS after 2011
		Start-up	
Columbia Generating Station (the Pacific Northwest's only active commercial nuclear power plant)		1984	Operated by Energy Northwest (formerly known as WPPSS)

	Key Dates	**Current Status**
3 massive plutonium separation plants, "canyons"; B, T, Redox	Built in Manhattan/ AEC era	Decommissioned
Fast Flux Test Facility (FFTF)	Built early 1970s	Cold standby/ decommissioning
Feed Materials Evaluation Facility (FMEF)	Built for FFTF	Cold standby
Plutonium Finishing Plant (PFP)	Ceased operation 1987	Decommissioned
PUREX Plant (Plutonium and Uranium Recovery by EXtraction)	Shutdown 1987	Decommissioned/ ISS equivalent
K-Basins		Large-scale removal/ stabilization
177 tanks (149 single-shell tanks; 28 double-shell tanks)		(8 SSTs fully retrieved) Continuing large-scale planning/stabilization/ cleanup*
Radioactive waste burial grounds		Hundreds of acres contaminated

* The treatment of tank waste will begin circa 2017 when DOE and Bechtel complete a $13 billion waste treatment plant complex now under construction.

All would have produced more weapons-useable plutonium—a material that presidential administrations of both political parties have agreed is in surplus.

So while the ship was turning slowly from its previous course to find new ways to deal with these and other technically complex challenges, DOE undertook more mundane tasks in order to show progress. The enactment of environmental cleanup legislation[2] by 1980 had spawned an industry based on removing and disposing of contaminated soil and treating groundwater. DOE invested heavily in these new "off-the-shelf" activities.[3]

This was by no means a bad thing. Cleanup of contaminated soils and poorly-disposed wastes prevented the further spread of contamination. Installing readily-available wastewater treatment systems and barriers to stop the spread of contaminated groundwater were sensible, overdue steps. These remedial actions allowed for the establishment of the Hanford Reach National Monument along the Columbia River and the Rocky Flats wildlife preserve to the benefit of surrounding communities.

By the early 1990s, it also had become clear that the "mortgage costs" of maintaining facilities such as Hanford's PUREX plant in an operable state were high. The PUREX facility was deactivated—with hazardous materials and wastes removed, interior contamination sealed, and utilities disconnected—by 1997, saving some $45 million per year that had been required to maintain the plant.[4] Also at Hanford, the more technically challenging tasks of removing spent fuel from the K-Basins and stabilizing plutonium materials at the Plutonium Finishing Plant took hundreds of millions of dollars and were not completed until relatively recently.

At sites where DOE contemplated its continuing fissile materials missions, such as the Y-12 uranium enrichment plant at Oak Ridge and the facilities for reprocessing irradiated fuel at Savannah River, the high mortgage costs continued.[5] In 2006, the National Nuclear Security Administration laid out a plan for greatly reducing the size of the Y-12 complex and consolidating plutonium work at a single site, all of which would reduce maintenance and security costs.[6]

Meanwhile, DOE began to deal with even more technically complex problems, many that also had controversial social and political dimensions as discussed elsewhere in this book. Some of the most significant of these will be highlighted here.

In a number of situations, entirely new facilities have been needed to deal with waste legacy problems. The challenge is to create and operate these facilities in a way that does not replicate, in 40 or 50 years, much the same kind of dismal situation that DOE faced at the end of the Cold War. This is particularly relevant to new treatment facilities for high level wastes at Savannah River, the Idaho National Laboratory, and Hanford, and for Oak Ridge's remote-handled transuranic wastes. Can DOE build, operate, and then close these, as well as new facilities to dispose of surplus plutonium, in ways that minimize collateral waste, potential long-term contamination, and the need for never-ending active maintenance?

DOE's contracting and management approaches, however, tend to exacerbate these challenges. To date, moreover, the department's systemic difficulties in accurately projecting and meeting budgets and schedules for major projects have sapped much of the confidence in DOE's ability to get the job done.

Demolition at the D-Reactor facility, a part of Hanford's decontamination and decommissioning activities, August 30, 2001.

During D-Reactor's Safe Storage Enclosure (SSE), Hanford workers make preparations to seal openings, December 31, 2002.

Some key policy decisions remain regarding how to treat high level waste. DOE has yet to determine the final form of—and thus the kinds of facilities needed to treat—a significant portion of Hanford's tank waste, as well as calcine waste in Idaho. The department is considering alternative treatments for Savannah River's spent fuel before it is sent for final disposal. These decisions open up politically sensitive issues about what is disposed of near-surface at the present sites, and what is transported to—and through—someone else's backyard.

These decisions depend on locating, developing, and opening the national repository for high level waste. This socially and politically complex problem has to be addressed in order to move ahead with the challenges of high level waste treatment. I will later suggest some steps that may help the country move through this morass.

Another complex technical issue is: How should we deal with contamination deep in the soil and groundwater? Access to it is extremely problematical. This contamination is beyond the reach of such conventional remedies as "muck and truck," capping, or constructing barrier walls. Often the problem also is politically complicated by the definition of contaminants: Should they be considered high level or transuranic waste, thus requiring removal and disposal in deep geologic repositories?

Avoiding Past Mistakes

As noted, the cleanup of legacy wastes and the disposition of plutonium (and additionally, of materials from ongoing nuclear weapons activities) will require new facilities and the continued chemical processing of nuclear materials. In building these facilities and conducting operations, DOE must avoid the past practices that have led to continuing problems.

From the very beginning in planning for new facilities and operations, people and organizations need to look ahead to waste management, and to a facility's eventual decommissioning, closure, and removal, as well as any continuing post-closure care requirements. The long-term State and Tribal Government Working Group, which was established to advise DOE's Environmental Management program, made the following recommendation in 1999:

> For new facilities and missions DOE should address the closure and long-term stewardship commitments associated with the facility or mission in the initial approval decision, and make provisions for funding of the closure and post closure operation of the new facility or mission.[7]

In its 2000 response, DOE acknowledged that its policies and orders did not provide clear guidance as to how to address these considerations in planning new facilities and operations.[8] Subsequently, DOE issued Order 450.1, establishing directives for Environmental Management Systems at its facilities and sites.[9] The order requires a more directed approach toward minimizing pollution, environmental monitoring to assure that regulatory requirements are met or exceeded, and consideration of life-cycle impacts of facilities and activities. It also includes a list of requirements to be passed on to DOE's contractors.

This order places some measure of accountability on DOE's decision-makers to avoid repeating past mistakes. However, its meaningful implementation faces significant obstacles. First of all, both DOE and its contractors naturally are mission-focused. At its simplest, if Congress has authorized and appropriated money to build a facility or carry out a program, the principal measure of achievement will be the accomplishment of that specific mission.

Second, DOE's approach to achieving its goals through contracting reinforces a near-term focus. Generally, the contractor selected for constructing a facility does not expect to be responsible for operating, decommissioning, or closing it. And, generally, other contractors operating a facility will not expect to be directly responsible when it comes time for decommissioning and closing.

A highly specialized team wearing full protective clothing and using respiratory devices enters the 242-Z facility for the first time in 16 years to make an assessment, June 2, 2005. One of Hanford's most serious contamination accidents occurred here on August 30, 1976.

Also, pressure for cost-effective performance makes these contractors more likely to turn their thinking toward near-term goals, rather than to long-term considerations. DOE's recent history of emphasizing incentives or bonuses for contractors who accelerate work within budget constraints may actually intensify this near-term focus.

Finally, DOE, as the owner of a facility and ultimately the responsible party, does not face the same financial assurance requirements that private owners of similar facilities must face.[10] Therefore, a future Presidential administration and future Congressional appropriations will be expected to deal with the long-term liability issue.

Under these circumstances, it takes a strong commitment on the part of DOE, the federal government as a whole, regulators, and the public to make sure that the long-term consequences of today's projects are seriously addressed. This is one reason why many agencies, individuals, and groups expend great amounts of energy reviewing and commenting on EIS and other assessments that the Department of Energy prepares pursuant to the National Environmental Policy Act.

Accurate Cost and Schedule Performance

Some of these contracting problems contribute to, and have been exacerbated by, DOE's own continuing difficulties in accurately forecasting costs and schedules for major projects. The Bush administration deferred construction of a projected facility at Savannah River to prepare surplus plutonium for vitrification because of escalating costs and scheduling problems for parts of the plutonium disposition program.[11] The removal and stabilization of Hanford's spent nuclear fuel inventory, while finally successful, also suffered from cost escalation and schedule slippage for most of a decade.

The Government Accountability Office has documented continuing cost and scheduling performance shortcomings at DOE for more than a decade. This was a period extending over two Presidential and five secretarial regimes, all of whom attempted to introduce more effective management and contracting approaches.[12]

The following discussion about the retrieval and treatment of high level wastes underscores the fact that these difficult, systemic problems have yet to be overcome.

Treating High Level Waste

Treating high level radioactive wastes—including the residuals from the reprocessing of spent fuel in order to retrieve weapons-grade plutonium and usable uranium—remains one of the most difficult and costly challenges of the post-Cold War cleanup. These wastes exist primarily at three sites—Hanford, Savannah River, and the Idaho National Laboratory.[13] At Hanford and Savannah River, the waste consists of nearly 97 million gallons of liquids, sludge, and salts contained within aging tanks. About 1 million gallons of liquid wastes remain at the Idaho National Laboratory's tank farm; the balance of INL's high level waste, about 4,400 cubic meters, is stored as calcined powder.

Since the 1980s, the basic approach to handling the liquid wastes stored in tanks at Hanford and Savannah River—the greater majority of the nation's high level waste volume—is to separate it into high level and low-activity fractions. The former will be vitrified into glass for disposal at a deep geologic repository. The low-activity fraction, on the other hand, is to be solidified in some form and disposed off on-site; at Hanford, the plans for the exact form in which this solidified waste is expected to be treated has varied, and continues to vary, over time. The current expectation is that the low-activity waste also will be vitrified into glass.[14] At Savannah River, the low-activity fraction is largely composed of precipitated salts, which are disposed in a cement-like grout called "saltstone."

A vitrification plant, the Defense Waste Processing Facility (DWPF), was completed at Savannah River in 1996. As of 2006, the plant had produced 1,969 canisters of glass, which is about 39 percent of the expected total. DOE built the Savannah River plant first for several reasons. Included among these was the fact that Savannah's liquid wastes were smaller in volume and more chemically homogeneous than at Hanford.

As with any large, complex, one-of-a-kind facility, the DWPF took more time and money to construct than originally estimated. Its operations have not been trouble-free either; waste treatment has taken longer and has cost more than projected. The original approach to separating high level wastes from low-activity salts by in-tank precipitation did not succeed. Thus, DOE was forced to revise its strategy and to develop new treatment facilities to try to maintain the momentum of vitrification at the DWPF.[15]

DOE's intent was that DWPF would be a prototype for a similar plant at Hanford, recognizing, however, that the volume and chemical complexity of the waste in Washington would require additional capabilities. DOE also assumed that, as the capital costs for Savannah River's DWPF construction ramped down, the means for the establishment of a vitrification plant at Hanford would increase.

Neither expectation proved accurate. Plans at Hanford have gone through four distinct reiterations, each moving away from replicating Savannah River's DWPF. Actual capital construction did not begin until 2001—a laboratory, a facility to separate high level and

Hanford's High-Level Waste Vitrification Facility under construction, January 31, 2004.

low-activity waste fractions, and plants to vitrify each fraction. The proposed site will handle four times the volume of waste as Savannah River's DWPF. The start of vitrification, however, originally predicted in 1989 to begin in 1999, is now projected for 2018. The estimated cost for the waste separation and vitrification facilities rose from $4.3 billion in 1996 to $11.3 billion a decade later.[16] (The later financial estimate was due to events in 2005. At that time, DOE had to halt construction so that several reviews of cost and technical issues, such as seismic stability and the adequacy of piping systems for processing, could be conducted.)[17]

The long delay at Hanford poses a significant dilemma. The wastes to be retrieved and treated are contained in 177 large underground tanks. Of these, 149 single-shell tanks (SSTs) were built between 1944 and 1970, and 77 of them have leaked in the past. The double-shell tanks, 28 of them, were constructed in the 1970s and early 1980s, with a nominal design life of 30 years. If current schedules are realized, the first tanks will be more than 70 years old, and the newest more than 30, before wastes from them can be treated.

The planned approach has been to move wastes from the older, leak-prone, single-shell tanks to the newer double-shell tanks. However, the dilemma looks something like this:

- Without the ability to treat the wastes already present in the double-shell tanks, DOE will have little room to store wastes from the oldest tanks.
- Should new, safer tanks be built to receive waste from the older tanks, resources and attention will be diverted from building and operating the waste treatment plant. Overall costs for the eventual managing and cleanup of wastes (which are in part a function of the greater amount of time involved) will increase.
- If new tanks are not built, the probabilities of additional leakage of waste from the existing tanks will grow at an accelerating rate. Additional leaks into the soil increase the long-term threat of contaminants reaching groundwater and the Columbia River. Intercepting, retrieving, and treating leaked waste is more difficult and costly than dealing with wastes still contained in tanks.

Hanford workers removing drums containing suspect transuranic waste from a burial trench. By mid-March 2004, more than 1,600 drums had been retrieved.

How DOE will handle Idaho's high level waste also is an open question. In early 2006, DOE decided to treat liquid wastes remaining in the INL tank farm by "steam reforming"—a process that will produce a granular powder similar to Idaho's existing calcined waste.[18] Because most of the remaining INL liquid waste did not come directly from reprocessing activities, steam reformed wastes may ultimately be approved for disposal as transuranic wastes at the Waste Isolation Pilot Plant. Alternatively, it could be sent, along with the calcined waste, to the high level waste repository. Whether either of these types of waste will be vitrified before disposal has not been determined.

In whatever form, the Idaho Settlement Agreement specifies that all high level waste must be ready for shipment from Idaho by 2035.[19]

Dealing with Deep Contaminants

For many years, the producers of nuclear weapons materials discharged liquid wastes into the ground. Evidence at hand at the time suggested that most of the biologically harmful radionuclides would be bound up in the soil—at least at arid sites such as Hanford and the Idaho National Laboratory. Furthermore, many of the more energetic elements would decay away in the arid soil before ever potentially reaching groundwater. (A state official once described Hanford to me as: "Hundreds of square miles of kitty litter.")

Officials at Hanford realized early on, however, that disposal into open ditches or ponds was a poor solution. Wind, migrating birds, and other wildlife could quickly spread surface contamination. Officials therefore decided on a safer approach for the liquid wastes they deemed fairly "hot," but not to the degree as to require storage in tanks. They developed cribs—underground drain fields similar to domestic septic systems.[20] Thus, liquid wastes were discharged into gravel-filled underground boxes.

These boxes held the waste for a time, allowing shorter-lived radionuclides to decay. Eventually the liquid seeped below into the earth, where it was thought that most of the remaining contaminants would be bound up in the soil. Over time, radiation protection officials installed monitoring wells to help determine whether contaminants were migrating outward more than expected.[21]

Scientists did recognize that uranium and more mobile contaminants were moving downward into the groundwater, and that there

Retrieval structure over Pit 4 at the Idaho National Laboratory, February 8, 2005.

were significant variations in the soils and geologic structure under Hanford's processing areas. They also realized that the discharge of large volumes of wastewater into the ground had created groundwater mounds, which affected the direction and speed of contaminant movement.

As a result of discharges into a particular set of Hanford cribs, such mobile contaminants as technetium-99, uranium, and iodine-129 not only reached the water table, but moved deep down. Some contaminants now are found 200 feet or more below the surface. The usual methods of dealing with soil contamination—e.g., removal and re-disposal, or isolation by capping or installing underground barriers—are not necessarily applicable in such a situation.[22]

One estimate of the cost to remove and dispose of contaminated soil below these cribs is nearly $1.7 billion.[23] Scientists admit the extent of contamination is not precisely known, nor are all of the factors of soil chemistry and geology that influence where and how contaminants move, and how quickly. Obviously, though, a portion of Hanford's wastes have traveled deep below the ground surface.

The early use of waste injection wells at Oak Ridge and Hanford, discharges of contaminated liquids at INL, and underground weapons blasts at the Nevada Test Site[24] all have led to deep pockets of contamination. In a number of cases, the wastes now residing deep underground contain transuranic elements, and, if exhumed, might meet the requirements for disposal at WIPP.[25] Some would argue that wastes leaked from Hanford tanks legally must be retrieved and disposed of at a geologic repository.

In short, a significant policy debate remains regarding these deep pockets of contamination. Key to the debate, of course, is whether or not it is technically feasible to retrieve these wastes, or, on the other hand, whether there are technically effective means to isolate the wastes in situ over a long period of time.

Finding a Repository

The nation's present approach to cleanup assumes that spent fuel and high level waste from Hanford, Idaho, and Savannah River will be joined with spent commercial fuel in a deep geologic repository.

As noted in the discussion about Nevada's Yucca Mountain site in Chapter 6, the development of a high level waste repository is at best 20 years behind schedule, and achieving the newly-revised projected opening date of 2018 is far from certain.

The broadly-based political and scientific agreement on a deep geologic disposal site for spent fuel and high level waste, arrived at by the early 1980s, is a good solution. However, DOE's efforts to implement this plan have been less than effective. The repository program has foundered for three principal reasons:[26]

- The discussion of risks involved at the repository and in long-term storage at existing sites has been cast in off-putting technical jargon. This has discouraged dialogue between scientists and concerned lay citizens. The jargon, moreover, often appears to be employed to justify or defend predetermined decisions, not to set forth in open discussion what is and is not known.
- The process delineated in the Nuclear Waste Policy Act relies on a series of rule-based "go, no-go" decisions. This approach is not well suited to a situation in which there are large uncertainties.
- The overly-optimistic timetable for repository development did not allow for step-by-step, incremental learning, or the development of common understandings along the way.

As the National Research Council Committee on Disposition of High-Level Radioactive Waste explained:

> **Today the biggest challenges to waste disposition are societal.** Difficulties in achieving public support have been seriously underestimated in the past, and opportunities to increase public involvement and to gain public trust have been missed. Most countries have made major changes in their approach…to address the recognized societal challenges. Such changes include initiating decision processes that maintain choice and that are open, transparent, and collaborative with independent scientists, critics and members of the public.[27]

While it is possible that the continued exercise of political muscle may ultimately prevail (e.g., in reference to the "screw Nevada bill"), I believe that a revised approach, incorporating some of the NRC committee's insights, will prove more viable in the long run. At a minimum, such an approach would include:

- Use of language that allows a wide range of participants to understand scientific information and uncertainties, and to express local and regional knowledge and concerns.
- A need for a structured dialogue, using Web-site based tools, among concerned individuals and groups regarding where high level waste and spent fuel are presently stored, repository site issues, and likely transportation routes for waste materials. An organization independent of pro- or anti-repository factions could manage such a dialogue. An expanded Nuclear Waste Technical Review Board [28] could periodically evaluate the process.
- A flexible timetable and an approach that focuses on step-wise learning to increase common understanding, reduce uncertainty, and build confidence when moving forward.

What Will It Take?

These challenges cannot be successfully met unless three interlinked conditions are met. The first is sustained political and governmental support, including the willingness to continue significant funding for cleanup over the long haul. This willingness will need to accept the pursuit of technological alternatives, some of which may not work out.

Second, such political support will not be forthcoming unless both the attentive public and their elected representatives trust those in charge of the cleanup. Such trust is hard to win and easy to forfeit. To be blunt, DOE has experienced failures in securing and holding the trust of the public or Congress in its cleanup program.

Third, openness, accountability, and public engagement are essential for building and sustaining trust.

In the final chapter, I will address the prerequisites for a sustained, effective response to these challenges. A cycle of accountability, openness, and public trust is at the heart of an ultimately successful cleanup of the nation's nuclear weapons production facilities. Without these elements, the success of any adopted cleanup program might only be minimal in the long run.

Notes

1. One senior official in the Clinton administration told a colleague of mine: "The weapons culture was so ingrained that, in turning the ship, we had to take the rudder out of the water and start beating the crew over the head with it."
2. The Resource Conservation and Recovery Act (RCRA), the Comprehensive Environmental Response, Compensation and Liability Act (CERCLA), and the Uranium Mill Tailing Remediation Control Act (UMTRCA).
3. *Complex Cleanup: The Environmental Legacy of Nuclear Weapons Production* (Washington, D.C., Congressional Office of Technology Assessment, February 1991), chapter 2.
4. Michele Stenehjem Gerber, *On the Home Front: The Cold War Legacy of the Hanford Nuclear Site* (Lincoln: University of Nebraska Press, 1992, 2002), p. 221.
5. At the time of this writing, deactivation of Savannah River's F-Canyon is finally underway.
6. Statement by Thomas P. D'Agostino, deputy administrator for defense programs, National Nuclear Security Administration, before the House Armed Services Subcommittee on Strategic Forces, April 5, 2006.
7. *Closure for the Seventh Generation: A Report from the Stewardship Committee of the State and Tribal Government Working Group* (Washington, D.C.: National Conference of State Legislatures, February 1999).
8. USDOE, *Draft Long-term Stewardship Study* (October 2000), chapter 6.
9. Order 450.1 was issued in January 2003 to implement a series of Executive Orders on "Greening the Government." Program offices and sites were required to have environmental management systems in place by the end of 2005. See "Order 450.1" at: www.directives.doe.gov/pdfs/doe/doetext/neword/450/o4501admc1.pdf.
10. The financial assurance issue was most clearly drawn when the State of New Mexico attempted to insert such a requirement, pursuant to the Resource Conservation and Recovery Act, into its WIPP permit. DOE asserted sovereign immunity, however, and Congress, in effect, concurred in the Military Construction Appropriations Act (2001): "Funds appropriated in this or any other Act and hereafter may not be used to pay on behalf of the United States or a contractor or subcontractor of the United States for posting a bond or fulfilling any other financial responsibility requirement relating to closure or post-closure care and monitoring of the Waste Isolation Pilot Plant. The State of New Mexico or any other entity may not enforce against the United States or a contractor or subcontractor of the United States, in this or any other fiscal year, a requirement to post bond or any other financial responsibility requirement relating to closure or post-closure care and monitoring of the Waste Isolation Pilot Plant. Any financial responsibility requirement in a permit or license for the Waste Isolation Pilot Plant on the date of the enactment of this section may not be enforced against the United States or its contractors or subcontractors at the Plant." Public Law No. 106-246, 114 Stat. at 536.

11. USDOE, "Surplus Plutonium Disposition Program: Amended Record of Decision," *Federal Register*, Vol. 67, No. 76, January 26, 2001, pp. 19432–35.

12. U.S. Government Accountability Office (GAO), *Nuclear Waste: Better Performance Reporting Needed to Assess DOE's Ability to Assess the Goals of the Accelerated Cleanup Program* (Washington, D.C., July 2005). GAO reports can be accessed at: www.gao.gov.

13. A relatively small amount of spent fuel waste remained at the West Valley site in New York. This was the one place where the reprocessing of commercial spent fuel to make new fuel was undertaken in the 1970s. Under the West Valley Demonstration Project Act (1980), DOE took over treatment of this waste, which was retrieved from storage tanks and vitrified into glass by 2002.

14. There is a larger volume of low level waste, but its handling requires less stringent specifications than for high level wastes. Studies of bulk vitrification (i.e., pouring glass into large, metal-box shipping containers) are underway.

15. *Savannah River Site Salt Processing Alternatives Final Supplemental Environmental Impact Statement* (SPA SEIS), June 2001. "Amended Record of Decision: Savannah River Site Salt Processing Alternatives," *Federal Register*, Vol. 71, No. 15, January 24, 2006.

16. The $4.3 billion and $11.3 billion figures, in each case, are for a comparable facility. The initial cost for a plant to treat only double-shell tank waste, the starting point in 1989, was about $1 billion. I am grateful to Suzanne Dahl of the Washington State Department of Ecology for summarizing this history in a presentation to the Oregon Hanford Cleanup Board, July 26, 2006. In January 2007, DOE approved a new baseline cost estimate of $12.26 billion.

17. U.S. Government Accountability Office, *Hanford Waste Treatment Plant: Contractor and DOE Management Problems Have Lead to Higher Costs, Construction Delays, and Safety Concerns* (Washington, D.C., April 6, 2006).

18. "Record of Decision for the Idaho High-Level Waste and Facilities Disposition Final Environmental Impact Statement," *Federal Register* Vol. 70, No. 242, December 19, 2005. See also, *Idaho High-Level Waste & Facilities Disposition, Final Environmental Impact Statement*, September 2002.

19. At the time of this writing, some members of Congress and federal administrators are suggesting that the reprocessing of spent nuclear fuel should once again be considered as part of the nation's energy policy. Reprocessing for commercial power generation purposes was abandoned in the 1970s, in large part because it was not economically viable; the experience at West Valley also indicated significant shortcomings in managing wastes and contamination. The proposed renewed reprocessing, of course, would produce more high level liquid wastes.

20. Roy E. Gephart describes the nature, evolution, and use of cribs in *Hanford: A Conversation about Nuclear Waste and Cleanup* (Columbus: Battelle Press, 2003), section 5.6.

21. There was no early consensus that soil absorption would work as anticipated. See the discussion in Michele Stenehjem Gerber, *On the Home Front: The Cold War Legacy of the Hanford Nuclear Site* (Lincoln: University of Nebraska Press, 1992, 2002), pp. 147 ff.

22. The complexities of dealing with these problems are presented in DOE Hanford Site, "Evaluation of Vadose Zone Treatment Technologies to Immobilize Technetium-99," August 31, 2005.

23. Undated graphic, titled "B/C Cribs—End State Alternatives."

24. The undersurface contamination at two Nevada locations are estimated to be at the 3,200 and 1,200 foot levels respectively. See documents related to the Central Nevada Test Area and Project Shoal Site Area at: ndep.nv.gov/shoal/shoal1.htm.

25. For an in-depth survey of transuranic ground contamination, see Marc Fioravanti and Arjun Makhijani, *Containing the Cold War Mess: Restructuring the Environmental Management of the U.S. Nuclear Weapons Complex* (Washington, D.C.: Institute for Energy and Environmental Research, 1997), chapter 2, "Transuranic Waste Management."

26. I developed this analysis in a paper, "'Substantial Margin of Safety': A New Approach to HLW Disposition," presented at Waste Management '02, Tucson, Arizona, February 2002.

27. National Research Council, Committee on Disposition of High-Level Radioactive Waste through Geological Isolation, *Disposition of High-Level Waste and Spent Nuclear Fuel* (Washington, D.C.: National Academy Press, 2001).

28. The Nuclear Waste Technical Review Board could be expanded to include publicly-nominated experts in risk communication.

9

Trust and Momentum to Get the Job Done

I n a thoughtful presentation given in 2000 about the defense nuclear cleanup program, Katherine Probst and Adam Lowe of Resources for the Future asked: "Does anybody care?"[1] Their main point was this: The cleanup program has little visibility "within the beltway"—i.e., among decision-makers and interest groups in Washington, D.C. This lack of engagement has led to inefficiencies and an inconsistent approach in the cleanup effort on DOE's part.

I will return later to some of the political factors that Probst and Lowe believe account for this lack of visibility in the Beltway. First, however, I want to explain how more openness, and consequently greater accountability, would make the cleanup program more effective and provide a base for continued public support. Indeed, openness and trust are factors that may become even more essential as the nation's geographical base for engagement in the cleanup effort shrinks.

At the Main 234-5Z Building in Hanford's Plutonium Finishing Plant (PFP), one-third of the 189 gloveboxes and hoods have been cleaned out, and 44 decontaminated to low level waste status, May 31, 2005.

Building trust and genuine visibility—a sharing of information and a willingness to listen—is key to sustaining the support needed to resolve the remaining complex, long-running, and expensive cleanup challenges.

Ultimately, the wisdom and sustainability of actions taken in response to the remaining cleanup problems at sites with residual contamination depend on the engagement of ordinary people. Developing the openness, trust, and accountability required for success must start among those residing around the contaminated sites—workers, community leaders, and people in neighborhoods. It would be too simplistic to say that this will necessarily translate into greatly increased accountability and support inside the Beltway, but the absence of these efforts at the local and regional level will certainly be draining on national support for cleanup.

The question is not so much one of passing along seriously contaminated sites to future generations (which in certain places is unavoidable), but rather it is one of doing so in a way that will make things better, not worse, over time.

Maintaining Openness in an Insecure World[2]

Openness on the part of government has been, and is, critical to cleanup success—particularly in regard to the government's willingness to consider a wide range of public concerns and values, and for allowing public access to information. Facts are the basis for both citizen involvement in decision-making and official accountability. The establishment of the Waste Isolation Pilot Plant and the closure of Rocky Flats would not have been achieved without this openness.

This could not be more clearly stated than by the Openness Panel of the Secretary of Energy Advisory Board in a 1997 report:[3]

> In coming years, DOE must carry out major responsibilities involving nuclear weapons, the management of radioactive materials, and environmental remediation. These will require DOE to select new facilities for producing radioactive tritium (so as to maintain the viability of the current inventory of nuclear weapons), and other facilities for processing radioactive wastes and surplus fissile materials and for storing and finally disposing of these materials. In addition,

Waste Receiving and Processing (WRAP) Facility at Hanford, August 1, 2005.

DOE will have to transport radioactive materials through many communities throughout the country. Given the high level of public concern and sensitivity about radioactive materials, and the continuing debate about nuclear weapons and nuclear power, these would be challenging tasks in the best of circumstances. The difficulty will be aggravated if the Department is suspected of hiding risks and of concealing past accidents. Openness—and the enhanced credibility that can come from it—is a necessary condition for success in these activities.

The Openness Panel focused on three areas—classification, accessibility, and culture:

Classification and restricted information. The panel recognized that strides had been made in the 1990s toward dealing sensibly with classified and restricted information. Indeed, it is important to safeguard certain information about the specific manufacture and deployment of nuclear weapons. However, the historic practice has been that all documents originating in the course of weapons production were assumed restricted unless specifically reviewed under complex rules for declassification; essentially, the documents were "born classified."

In the 1990s, the panel and DOE moved toward a different approach, requiring that classification or restriction of new documents require an affirmative decision. They also proposed segregating sensitive information in appendices or specific sections of documents, so that the balance need not be classified. At the same time, DOE moved rapidly to review and declassify historic documents and photos.[4]

Accessibility. Documents that are not restricted or have been declassified may nevertheless remain effectively unavailable to an interested party. DOE's historical approach to records management has compounded the problem. Most records were generated and held by field offices, not at headquarters, and were managed by contractors. Often contractors that were replaced at a site took their records and data with them.

The shift from paper to electronic records also has complicated matters. Electronic formats have evolved rapidly, and earlier formats may be difficult to retrieve or search. On the other hand, DOE made significant strides in building databases and scanning historic paper records to improve accessibility through the World Wide Web.[5]

However, after the terrorist attacks of September 11, 2001, on the World Trade Center and the Pentagon, DOE withdrew some information (e.g., EIS and other assessments containing detailed site maps or drawings of nuclear facilities) from Web site access. Generally, however, these documents are still publicly available in paper form at designated libraries.

Culture. A personal experience I had illustrates the Openness Panel's view of new strides regarding "culture." One afternoon in February 1986, a colleague and I happened to be present at the Federal Building in Richland, Washington, when Steve Leroy, the director of communications for the Richland Operations Office, invited us to sit in on a meeting with senior contractor officials. It was the eve of the release of 19,000 pages of previously classified or restricted documents that revealed the extent of the historic releases of radioactive materials from Hanford operations.[6]

The message in Steve Leroy's presentation was simple: What was done was done. We don't have to apologize for it. We have nothing to hide. This is a different time.

His audience, however, was hardly comfortable with this shift to openness. But Leroy's boss, Richland Manager Mike Lawrence, understood that continued secrecy would make it increasingly difficult for DOE to carry on its nuclear activities.

This illustrated the point later stressed by the Openness Panel in advocating a culture change. Mike Lawrence showed leadership, diverted resources from current missions to declassify and make available thousands of documents, and made his expectations clear to both employees and contractors.

Based on the Openness Panel's work—which stressed the need for a continuing emphasis on declassification and accessibility—Secretary of Energy Federico Peña publicly released a large volume of documents describing the nation's plutonium production and holdings.[7]

However, official attitudes and accessibility to information have retrograded since 9/11. The Secretary of Energy Advisory Board and its Openness Advisory Panel were disbanded in May 2006. While the affirmative-decision approach to classification adopted in the 1990s continues, the terms under which officials may now justify classification and restricting measures have been broadened.[8]

The Homeland Security Act (2002) created a new exemption from the Freedom of Information Act (1966); this restrictive category is called Critical Infrastructure Information (CII). According to recent reports, federal officials now can deem information as sensitive, though not meeting criteria for classification, and presume to deny public requests under the Freedom of Information Act.[9] And, the relatively strict rules covering classification[10] do not govern other exemptions, leaving wide latitude for discretion by individual agencies and officials.

On December 14, 2005, President Bush issued Executive Order 13392, "Improving Agency Disclosure of Information." This requires agencies to prepare responsive Freedom of Information Act plans that will be reviewed by the Office of Management and Budget and the Attorney General. These actions may help develop more consistency, but will not necessarily lead to more openness.[11]

In the public engagement that is critical to making future progress in dealing with nuclear waste issues, the dialogue should include a determination of a balance between government's "securing" of sensitive information, while also providing data needed for public

Downwind and Downriver

Potential Public Harm

Working at defense nuclear facilities always has been risky. Even in the earliest years, the Manhattan Project and the Atomic Energy Commission worked hard to understand and protect workers from the dangers of radiation. In recent years, the government and contractors have adopted much better overall safety practices at these sites. Moreover, one can argue that workers were compensated by relatively high wages, and they accepted the challenge of dangerous work in the cause of national defense.

But what about the folks living in rural areas and communities near the sites—many who had no real idea of AEC missions? In the early years, for example, chemical reprocessing of irradiated fuel at Hanford to recover plutonium resulted in the emission of large amounts of iodine-131 in the environment and into the prevailing winds out of the southwest. Iodine-131, if inhaled or ingested, lodges in the thyroid, where it may cause tissue damage. The half-life of I-131 is not long—about 8 days. But that is enough time for it to land on grass, be consumed by dairy cows, and reach children and other consumers. By 1951, radioactive air emissions were reasonably well understood and finally controlled at Hanford and other facilities. But there certainly were accidental releases at defense production sites—e.g., from fires in Rocky Flats plutonium processing buildings in 1957 and 1969.

A Hanford worker dismantling a 66-inch effluent pipeline as part of the environmental restoration effort, January 31, 2003. The pipeline returned reactor cooling water to an outfall into the Columbia River.

Water-borne contamination is a different matter. Operations to produce fissile materials and attendant waste disposal practices created the potential for downstream harm. Hanford's first 8 nuclear reactors were "single-pass"—i.e., cooling the reactors required large amounts of Columbia River water, which, after some delay to allow radioactive contaminants to decay, was then reverted back into the river. As Cold-War tension mounted in the early 1960s, the government ran these reactors at higher power and with less hold-back time. Radioactive contamination of the lower Columbia River—the boundary between Washington and Oregon and numerous communities—peaked in the 1964–1966 period.

The extent of potential harm to people downriver is a complex question. Few municipal water systems or irrigators drew directly from the

L-Area oil and chemical basin at Savannah River, December 1, 1993.

Columbia in those years, and the very large flow of the river quickly diluted many contaminants to undetectable levels. Still, those who, like American Indians in the region, consumed resident bottom fish were at risk. Hanford's radioactive products were detectable in shellfish along the southern Washington and northern Oregon coast in the mid-1960s.

During the Manhattan Project and the Cold-War years, operators at most defense nuclear sites discharged contaminated liquid waste into the ground and/or disposed of solid waste in near-surface dumps. Such practices were particularly problematic at sites such as Oak Ridge and Savannah River, which are wet environments, with a lot of surface water, and the groundwater is near or at the surface.

At more arid sites, such as the Nevada Test Site, Hanford, and the Idaho National Laboratory, the movement of contaminants into the groundwater and then into wells or surface water may take longer, but such movement is harder to measure and predict. Suffice it to say, at least at Hanford, earlier government and contractor estimates consistently underestimated how much, and when, contaminants would move toward the Columbia River.

Since the end of the Cold War, the government has conducted many dose reconstruction and epidemiological studies around nuclear defense production sites. See the Web sites for the Centers for Disease Control Radiation Studies: www.cdc.gov/nceh/radiation. Also, the Agency for Toxic Substances and Disease Registry (ATSDR): www.atsdr.cdc.gov/2p-toxic-substances.htm.

understanding and acceptance of significant decisions. Specific clarity about what should and should not be disclosed, moreover, can only reduce friction and mistrust.

Dealing with the Perceived Human Costs

The dialogue about cleanup will not be complete, in the minds of many, until the nation has fully owned up to the negative health effects to workers and "down-winders" resulting from nuclear weapons production and waste management decisions. Along with openness, this is key to building and sustaining the trust necessary to resolve the tough cleanup challenges ahead.

This is not an easy task. Since the late 1980s, the government has carried out a number of health effects studies, using the available tools of epidemiology. As with many scientific efforts, these studies are often inconclusive, particularly as they might apply to specific individuals, as opposed to some hypothetical percentage of a larger population.[12]

Early studies dealt with uranium miners in the West and those downwind from atmospheric bomb tests, particularly in the Marshall Islands and southern Utah. Later, the Hanford Dose Reconstruction and Thyroid Disease studies focused on those populations residing downwind and downriver from the early plutonium production operations. Still later, the Agency for Toxic Substances and Disease Registry (ATSDR) and the Centers for Disease Control conducted studies around Hanford and other defense nuclear sites. The National Institute for Occupational Safety and Health also investigated links between nuclear worker exposures and diseases.

The good news: Congress progressively opened up avenues of compensation to uranium workers, those down-wind from early atmospheric tests, and eventually to workers at defense nuclear facilities. These laws generally operate on the principle that if someone was in a certain location at a certain time and has diagnosed health problems often associated with exposure, then the symptoms more likely than not resulted from that exposure. Most recently, Congress expanded the Energy Employees Occupational Illness Compensation

Program Act (EEOICPA) to include exposure to toxic chemicals as well as radiation.[13]

The bad news: Until quite recently, DOE and its contractors spent as much effort, or more, on legal means to fend off injury claims in the courts than on compensation.[14] In many cases, the claims may not have been valid and the legal defense against them was appropriate. However, the impression left with concerned observers, let alone the plaintiffs, has fed mistrust. Initial sluggishness in processing claims and providing benefits under EEOICPA has fueled skepticism about DOE's commitment to redress past harm.[15]

As cleanup progresses, DOE and its contractors are aware that worker health and safety concerns require more attention than they had received in the Cold War era. There are currently in place a number of programs directed at work safety, including the Integrated

Highly radioactive reactor fuel fragments mixed with fuel spacers discovered in a Hanford burial ground along the Columbia River, February 28, 2005.

Safety Management System (ISMS) under which workers are involved in the planning and preparation of work tasks.[16]

In actuality, cleanup activities often are more dangerous than routine production tasks have been. At cleanup sites, facilities are older, so that structural, mechanical, and electrical failures are more likely. Records and/or on-site knowledge about the materials and chemicals stored in old facilities may well be lost or incomplete. And, with time, contaminants alter chemically or move in unexpected ways.

Understandably, DOE has paid considerable attention to worker risks when evaluating alternative cleanup actions in various environmental impact reviews. These analyses often weigh and balance the near-term risks to workers in proposed cleanup projects, as juxtaposed to the long-term risks to surrounding populations if contaminants are not removed or isolated.

While this is an important consideration, DOE sometimes in a specific situation comes too quickly to the conclusion that the long-term risk is not sufficient to justify cleanup actions exposing workers to real and known health risks. As one senior Hanford official put it at a public meeting a few years ago: "The safest work is work we don't have to do." A number of workers at that same meeting, however, responded that they believed the cleanup work needed to be done, but that DOE and its contractors were not taking all the actions necessary to protect them while doing it.

In short, cleanup success depends both on actions that are perceived to deal honestly and fairly with past harm to workers and surrounding populations, and on dealing squarely with potential harm to workers doing the cleanup today and tomorrow.

Effective Long-term Stewardship

Local, tribal, and regulator skepticism regarding an unclear DOE commitment to long-term protectiveness and its minimizing of costs in near-term cleanup has been a continuing source of mistrust, especially regarding sites having no long-term nuclear mission. In a nutshell, the mistrust is a result of DOE's frequent inability to provide financial assurances, and its assertion of sovereign immunity with regard to state and local laws. These factors can drive the other parties to demand a complete cleanup, in which no residual or potentially problematic contamination remains; in many cases this is not a realistic expectation.

As Milton Russell of the Joint Institute for Energy and Environment has argued, the only truly effective way for DOE to implement cleanup solutions requiring long-term stewardship is to engage other parties as partners[17] rather than treating them as generally unreasonable and ill-informed. There are some creative ways to build partnerships that will keep alive knowledge of what has occurred at a site (e.g., what remedies have been implemented, and how these must be managed and monitored). DOE and the federal government generally need to pursue institutional arrangements with state, local, and tribal governments and organizations to assure that this knowledge is maintained and accessible, that the public is made aware of it, and that specific parties into the far future can be held accountable to maintain safeguards.

The creation of trust funds is one potential mechanism.[18] There are two particular advantages in maintaining trust funds. First, of course, money set aside earns interest, which will fund the monitoring and maintenance of the physical and institutional controls for long-term site monitoring and containment. Second, the trust fund structure depends on trustees—i.e., specific interested individuals with fiduciary responsibilities regarding the purposes and beneficiaries of the trust.

The beneficiaries can be various. For example, one party, or a set of parties, can establish and fund trusts and designate trustees to take care of residual contamination. Selected trustees can disperse funds for a variety of purposes, such as information management or field monitoring, on behalf of local beneficiaries over the very long term. Any of the beneficiaries would have legal recourse to ensure that a trustee meets the required obligations.

I noted earlier that the federal government cannot commit to future appropriations, thus it must find ways to set aside money from present appropriations to build future trust funds. This is difficult to do, because Congress generally expects its resources to be applied to present-day needs. Its own budgeting practices mean that most federal "trust funds" do not, in fact, have money set aside drawing interest.[19]

However, there have been some creative efforts to generate such funds. In one case, DOE agreed with the State of Tennessee to establish a perpetual maintenance fund for its on-site waste disposal cell at Oak Ridge. DOE agreed in good faith—reserving all its legal rights to step away from the agreement—to deposit a set amount each year for 10 years into the state-managed fund.[20]

At San Francisco's Presidio, the Department of the Army agreed to contract its cleanup activities (non-nuclear contaminants) to a nonprofit organization dedicated to preserving the area as a park. A part of the fee the Presidio Trust collected for conducting the cleanup was set aside in a perpetual-care trust fund.[21]

A colleague, John Price, and I have proposed an innovative approach for the Hanford Site.[22] Our plan would capitalize on four institutional factors at Hanford:

- The existence of the Hanford Reach National Monument, administered by the U.S. Fish and Wildlife Service.
- The formation of an active, community-based, B-Reactor Museum Association dedicated to preserving the B-Reactor, the world's first large-scale atomic pile and thus an international historically significant site.
- The strong interest among local governments in developing tourism.
- The involvement of American Indian tribes with both knowledge about the Hanford area and strong oral traditions.

In our view, the establishment of a visitor center and museum at the B-Reactor in the northern part of Hanford, or in the City of Richland to the south, or at both places, could form the basis for an active stewardship program. DOE could enter into an arrangement with a locally-based nonprofit group, or the U.S. Fish and Wildlife Service, or both, to transfer and manage records and to participate in endowing a trust fund.[23]

Thus, a museum and/or visitor center complex would become a repository for historic records and cleanup information. A permanent board would assume responsibility for conducting ongoing programs that, among other things in the many decades to come, would remind future generations about the cleanup activities undertaken, the

B-Reactor at Hanford, January 31, 1999. The operation-support structures that originally stood adjacent to the reactor building have been removed.

residual contamination, and ongoing actions to keep contaminants isolated. Such a facility could also serve as a research center for scholars studying the nation's nuclear history and its programs.[24]

Also, a nonprofit board or the U.S. Department of Interior could contract with tribal members to serve as interpreters of Hanford's natural and human history. Engaging tribal people has three inter-related advantages. First, it provides another layer of awareness and accountability—redundancy is desirable in itself. Second, tribal members have a strong oral history tradition; this becomes a resource for keeping alive and re-invigorating the story of Hanford through many generations. Third, as Russell Jim—long involved in these issues on behalf of the Yakama Indian Nation—often says: "After the Europeans have come along, messed up the environment, and moved on, our people will still be here." Tribal peoples are deeply attached to land and place.

Partnerships developed along these lines can produce several ben-efits. As already implied, the partnerships would build trust for the future, giving DOE more latitude to pursue cleanup measures that necessarily must leave some contamination in place. In this regard, widespread agreement on the Rocky Flats cleanup was achieved only when Congress established a wildlife refuge and assigned the site's administration to the U.S. Fish and Wildlife Service. This reduced public concern about the effectiveness of institutional controls and future accountability in regard to the contaminants that remained at Rocky Flats.

Also, the more participants engaged in stewardship, the more likely it will succeed, building social and organizational redundancy into long-term protective measures. The National Research Council's report on long-term protectiveness at DOE's contaminated sites stressed the importance of redundancy.[25]

The pursuit of a positive vision for the future by a wide range of interested parties is crucial. It will sustain political support for the remaining cleanup challenges at defense nuclear sites. Now let us turn to the problem of sustaining political momentum.

Maintaining Momentum over the Long Haul

Since the time when Probst and Lowe asked "Does anybody care?" DOE indeed has made considerable progress in cleaning up contaminated sites near large urban areas, including at Rocky Flats, Fernald and Mound in Ohio, and Weldon Spring, Missouri. And DOE has managed, as also noted earlier, to reduce a number of highly apparent threats elsewhere. At Hanford, spent fuel has been removed from water basins near the Columbia River. Plutonium residues also have been stabilized at Hanford and elsewhere. The disposal of Idaho's buried transuranic wastes met the legal deadlines. At Oak Ridge in Tennessee, the removal and re-disposal of buried wastes has reduced threats to local streams and rivers.

Paradoxically, as remedial action narrows the list of sites with near-term threats and other concerns, it may well sap the overall political will to deal with the remaining and most complex challenges facing the cleanup program. In 2000, Probst and Lowe suggested four reasons why DOE's Environmental Management program lacked the national attention and accountability needed to make it truly effective:

- The great complexity and scope of the remaining cleanup challenge.
- Most major facilities are located in remote or lightly populated areas, and thus away from politically influential large urban areas.
- Funding is through Defense appropriations, thus budgets receive relatively little attention in Congress.
- The jobs-sustaining attitudes of communities hosting nuclear defense facilities.

Despite the recent cleanup accomplishments, these observations may be even more relevant today than they were in 2000. If attention and accountability are requisites for effectiveness, then everyone concerned about cleanup must address some significant institutional realities. First, as already noted, the remaining problems are extremely complex and challenging. Solving them will require technological development, contracting approaches that bend established federal practices, and constituents who are patient, knowledgeable, and demanding, not of perfection, but of accountability.

With the successful cleanup of the sites near urban population centers in Colorado, Ohio, and Missouri, there no longer will be issues in those areas generating intense political involvement. Also, as localized problems are addressed at the national laboratories in the San Francisco Bay area and on Long Island, interest in cleanup likewise will fade among politicians from two other large, influential states. The remaining cleanup challenges are primarily in rural localities, away from large metropolitan centers. Congressional delegations from states such as Washington, South Carolina, Tennessee, and Idaho will face an uphill battle to secure broad and continued cleanup support. Their colleagues from other states and the media might more readily characterize continued large expenditures for cleanup as "make work" in communities that have grown dependent on the Department of Energy.[26]

Furthermore, events since 9/11 have increased the size of other Defense appropriations, in relation to those for cleanup, and also shifted Congressional attention to other areas covered in the Defense budget.

Efforts to Focus Public Attention on Cleanup

Barring a major nuclear catastrophe that can spur broadened public outcries for action, there appears to be only limited means available to help sustain focus and momentum in regard to cleanup. Two approaches focusing on individual sites have had some degree of success; however, neither has always been especially compelling or particularly visible.

First, there are the organized citizen advisory boards at major DOE nuclear sites. These boards were the result of a broad-based dialogue that was facilitated among many agencies and departments in the early 1990s.[27] The boards attempt to include representatives with a diverse set of interests in evaluating issues and the oversight of cleanup activities. These advisory boards, at their best, keep local interest groups engaged, and provide a way to build a consensus in moving forward. They also help focus regional media attention on cleanup issues.[28]

Second, an accepted positive vision for future uses and activities at a particular site can sustain focus and momentum, and, indeed, has been effective in a number of specific places. For some sites, such as Rocky Flats, Weldon Spring, and Fernald, this entailed the preservation of diverse natural/recreational landscapes in or near growing metropolitan areas. The fact that the adjacent metropolitan areas were growing—with good local employment alternatives—eased fears about the loss of employment opportunities, which otherwise can be a major obstacle to local acceptance of such future uses.

At other major sites, such as Oak Ridge, INL, Los Alamos, and Savannah River, there is fairly broad agreement on future plans related to continuing nuclear activities. These sites are located farther from major metropolitan areas and there are fewer employment alternatives. There is a sense that cleanup at these sites must be resolved expeditiously, in order to secure continuing nuclear missions. The surrounding communities view new and continuing nuclear programs, not cleanup, as their primary source of permanent jobs.

On the other hand, at Hanford and the Nevada Test Site, broad all-encompassing agreements for either vision—i.e., creating a natural preserve, or adopting new nuclear missions—have yet to be achieved.

Both sites, however, may be diverse and large enough (586 and 1,350 square miles respectively) to accommodate both goals. Still, the lack of a unifying vision has been making it harder to sustain cleanup focus and momentum.

A number of national groups also continue to maintain attention on cleanup issues. States are involved through the National Governors Association, and the National Conference of State Legislators. Affected Indian tribes participate in the National Congress of American Indians, and the Council of Energy Resource Tribes. However, consistent with Probst and Lowe's observations about low visibility "within the beltway," cleanup is not a central interest of these broader state or tribal organizations.

The states and affected tribes also participate in the State and Tribal Government Working Group, and the Transportation External Coordination Working Group, both organized and supported by DOE. Both are constructive forums; however, neither really has much standing with DOE's top management or Congressional overseers.

The Alliance for Nuclear Accountability is a loose organization of site-specific environmental and watchdog groups, whose members gather periodically in Washington, D.C., to share information and for Congressional lobbying. In regard to national environmental organizations, perhaps the Natural Resources Defense Council has been the most active and consistent advocate for cleanup.

In recent years, the Government Accountability Project has expanded its engagement beyond whistleblower protection to include broader cleanup issues.

The Energy Communities Alliance represents local governments in those localities hosting defense nuclear facilities. The Alliance maintains a Washington, D.C., office, and periodically gathers there to share information and make Congressional contacts.

None of these organizations, however, provides a particularly powerful, sustained focus for the overall cleanup program. They can be rather invisible and, at least with regard to nuclear site cleanup, hardly anyone holds them accountable.

While not particularly visible, these groups do, however, provide multiple channels through which ordinary lay persons and groups can ask questions, gain information, and influence those engaged with cleanup (see Figure 9.1). Unless ordinary citizens get engaged—holding accountable not only DOE's responsibility for cleanup, but also state regulators and others whose job it is to also be effectively involved—it is unlikely that sustainable, effective solutions will be found for a number of cleanup issues.

Leveraging Change

Some lay citizens and groups have, indeed, revealed local problems from nuclear site contamination, and then generated political and public momentum to address these issues. Here are two stories illustrating how it can be accomplished.

For nearly five years, Lisa Crawford,[29] her husband, and young son lived in a rented farmhouse across the road from the Feed Materials Production Center at Fernald, Ohio. They thought the big industrial complex, with its red-and-white checkerboard water tower, had something to do with agricultural feed. Then one day in 1984, the plant released a radioactive dust cloud over them and their neighbors. Lisa began to ask questions.

K-65 radon testing during the Fernald Closure, November 30, 1993.

Figure 9.1
Defense Nuclear Site Cleanup Internet Resources
(Nearly all of these Web sites also provide links to additional information sources)

Organization	Description	Web site
Alliance for Nuclear Accountability	Association of site-related activist groups	www.ananuclear.org
Energy Communities Alliance	Association of host community governments	www.energyca.org
National Governors Association—Federal Facilities Task Force	Representatives of host states	www.fftcleanupnews.org
State and Tribal Government Working Group	DOE-sponsored state and tribal group on cleanup	www.em.doe.gov/pages/stgwg.aspx
U.S. Department of Energy	Access to Environmental Management, National Nuclear Security Administration, Environment, Safety and Health and individual DOE office sites	www.energy.gov
USDOE's Central Internet Database	Information on wastes at DOE sites	cid.em.doe.gov
USDOE's OPENnet	Historic information on weapons production and contamination	www.osti.gov/opennet
Transportation Resource Exchange Center (T-Rex)	DOE-sponsored information on nuclear waste transportation	www.trex-center.org
Nevada Agency for Nuclear Projects	Nevada position on Yucca Mountain—but also comprehensive daily updates on nuclear developments	www.state.nv.us/nucwaste
Nuclear Information and Resource Service	Information to support activists	www.nirs.org
Nuclear Control Institute	Focus on non-proliferation	www.nci.org

Early the next year, the landlord called to tell the Crawfords that the Department of Energy was sampling their well. At a heavily attended public meeting, the Crawfords learned that three wells had tested positive for excess uranium—and theirs, it turned out, was one.

Lisa called the Ohio Environmental Protection Agency, which advised her to find an alternative source of drinking water. Soon, several neighbors, mostly women, formed a citizens' group, Fernald Residents for Environmental Safety and Health, or FRESH.

"We were a shoe-string organization, making it up as we went along," Lisa recalls. "We spent a year getting information out to the public. We had to educate ourselves. We didn't know what a picocurie was. It can be overwhelming."

The group's efforts, however, soon showed results.

"Fernald was on the front page of the paper nearly every day. There was no more 'I don't know what they do there.'"

The Crawfords filed suit against the government. Eventually the suit, which was settled out of court in 1989, became a class action effort. A good deal more about Fernald and its off-site impacts emerged in the course of interrogatories and new revelations.

Lisa was particularly disturbed when government attorneys responded to one interrogatory, saying in effect: "Yes, we knowingly released this [nuclear] material and the plaintiffs can't do anything about it."

"The community was very angry. The more we learned, the madder we got."

DOE ceased producing uranium metal at Fernald in 1989. In 1990, Lisa and other FRESH representatives met "on day one" with the new cleanup contractor to present their expectations, including a seat at the table. "[The contractor] didn't have the right people at first. They were the old bomb makers. After a couple of years, we weeded them out, and then the contractor wanted us at the table."

Early on, Fernald's union workers were vehemently opposed to FRESH. When the new cleanup contractor came to Fernald, however, FRESH supported the union's campaign to retain the existing work force and bargaining rights.

Dewatering Excavation Evaluation Program (DEEP) at the Fernald cleanup, September 8, 1994.

"We were a group of forceful women. Every two years, we were in the offices of our elected representatives right after election to say: 'Here's who we are and what we expect.'"

In the lead up to the 1992 elections, FRESH members contacted the Clinton-Gore campaign early on, again calling attention to their concerns and expectations.

"We brought the bomb makers to their knees," Lisa recalls. "We worked with people in the community and with our elected officials." When it came time for the government to make legally-binding cleanup decisions, "we had to be ready to comment, based on the documents. You have to know what you're talking about."

Cleanup work at Fernald was completed in the fall of 2006. FRESH held its last meeting in November of that year, but agreed to continue in existence until the transition of the site to DOE's Office of Legacy Management was completed.

"We have a good cleanup level. Fernald is now totally green space," Lisa observed. "We used a balanced approach. We had a lot of waste. At a public meeting, I raised the question whether we could really send it all to someone else's backyard. It wasn't a very popular question, but eventually we decided as a community to keep the majority of waste on-site in a disposal cell."

Proportionately, about 80% of Fernald's waste remains on-site; 20% was shipped to Nevada and Utah disposal sites.

"It doesn't happen quickly. It took 22 years. People get burned out, tired. Members die."

Lisa noted that none of the members of FRESH knew one another prior to 1985. "These relationships are important. They never would have happened otherwise."

She also stresses the importance of networking with people involved at other nuclear sites.

"It comes down to a trust issue. In the early years there was no trust—especially with the government."

* * *

Another grass-roots effort also sprang to life in 1984 some 2,000 miles west of Fernald. Shortly after the government restarted Hanford's PUREX plant, William Houff, a Unitarian minister in Spokane, Washington, delivered a sermon titled "The Silent Holocaust," calling attention to the dangers of plutonium production. Congregants who heard Houff's sermon were highly motivated to act on the issue, and reached out to others as they formed a study group.

Lynne Stembridge,[30] a part-time office worker and mother of two young children, joined the group. Lynne had started a local chapter of Women Against Nuclear Armament, but knew very little about Hanford. Soon, the study group grew into the Hanford Education Action League (HEAL), and Lynne became a board member and then executive director.

The daughter of an Air Force B-52 bomber pilot, Lynne described her family and herself as "very patriotic." But as she investigated information about Hanford's past and present activities, she realized that people nearby had not been told about the potential harm.

"I was really shocked that the government would be deceptive to its people. That more than anything got me hooked."

Lynne was determined that "the government can't behave that way."

When HEAL and others secured the release of historic documents about off-site, open-air, releases of radiation, Lynne's conservative and patriotic mother handed her the fee for HEAL membership, saying: "I couldn't believe you were right, but I see you were. You've got to keep up this work."

While many who joined HEAL opposed nuclear armament and/or nuclear power, the group decided that those types of stands would not gain much ground in conservative Spokane. Rather, their messages would be clearly focused on Hanford's activities: "If you are going to do this work, then you must do it using the best science and you must be protective of public health."

Through the efforts of gifted volunteers, a few of whom later became staff members,[31] HEAL produced well-researched documents showing that neither the test of "using the best science" or being "protective of public health" had been met by the Department of Energy.[32]

Lynne's assessment of the impact of HEAL's activities: "We shut down plutonium production at Hanford. I don't think cleanup would have happened without our efforts combined with those of others. We had the right people in the right place at the right time with the right information. We used our passion to leverage change in the world."

Asked about getting lay people engaged in technical issues, Lynne responded that there are certain basic values that everyone can understand.

"You have to show your work so others can see it. Be clear about your assumptions. Tell the truth. Support your viewpoint with facts. Without supporting facts, opinion even if spoken loudly is still just opinion."

With a background in English and history, she bridled at experts who believed that "if you're not an expert you're a ninny, so we can't tell you anything. You can't handle it, or understand it, or help fix it in the end."

When asked about the state of nuclear site cleanup today, Lynne responded: "Like any endeavor that takes a long time, it takes patience, dedication, and watchfulness. It's like raising a teenager with an extremely long adolescence. They [the government] won't clean up their mess completely or properly without a lot of oversight."

The Continuing Need for Public Engagement

The stories of FRESH and HEAL illustrate basic principles of citizen engagement with controversial and technically complex issues. Acting on these principles, both organizations leveraged major changes in the policies, directions, and attitudes of an entrenched bureaucracy.

Those changes, however, are not likely to be maintained without continued involvement by concerned citizens. George Washington Plunkitt, a 19th century leader in New York's infamous Tammany Hall political machine, said: "Reformers is only mornin' glories."[33] Critics of DOE and its cleanup program often refer to "weebees"—i.e., in reference to those bureaucrats and contractors who say: "We be here when new political appointees come in to direct us, and we'll be here when they leave."

Hopefully, *America's Nuclear Wastelands* has illustrated how difficult it can be to effect cultural change within the nuclear weapons production complex. The "weebees" have the advantage over the "mornin' glories." An ordinary citizen may feel insignificant when facing entrenched institutions and in understanding the technically complex cleanup issues. As Lynne Stembridge pointed out, however, if people bond together, conduct effective research, and present the facts, they can have a lot of mass and can move in their own inexorable way.

If these different inexorable forces do not engage one another with openness, mutual respect, and accountability, then nuclear site cleanup will be unsuccessful:

- Contaminants will not be contained in the long run, thus future generations will face even more challenging problems.
- Narrow, piecemeal technical solutions will not incorporate important community values.
- Trust between government and communities will deteriorate.
- Potentially positive public resources and support will be squandered.

The tools are there to achieve better outcomes. For example, rich sources of information and documentation are available, many of which I have cited in this book. As President James Madison noted: "A popular government, without popular information or the means of acquiring it, is but a prologue to a farce or tragedy or perhaps both." The members of FRESH and HEAL have ferreted out facts, and shared and used this information to leverage fundamental improvements in regional cleanup efforts. And today, even more information has become readily available for access.

A new generation of concerned citizens, using these resources with the same kind of tenacity, reason, and skill, can assure that the cleanup will make things better for host communities and future citizens. Let me suggest three conclusions to be drawn from the experiences of citizens who have become engaged in nuclear site cleanup issues:

- Have a focused message based on a reasonable and transparent analysis of the information at hand.
- Be prepared to engage in assertive, but respectful, dialogue with people holding a wide range of differing views and values.
- Do not lose sight of basic principles in the face of complexity.

Finally, it is important to recognize that many individuals within the Department of Energy, its contractor community, and the regulatory agencies understand the importance of openness, trust, and accountability. These men and women want to conduct as successful a cleanup as possible. They reach out to concerned citizens, expecting to engage with and to learn from them. Effective citizens' groups, such as HEAL and FRESH, in turn have sought them out, worked with them, and supported them in their efforts.

To the extent that citizens, officials, and experts share lessons from the past, the country's efforts to clean up "the Cold War Mess" will not be a prologue to ineffectiveness or possibly even tragedy. Rather, we, as a nation, will be better able to deal with this and other complex problems that arise with any burgeoning technological development.

Notes

1. Katherine N. Probst and Adam I. Lowe, *Cleaning Up the Nuclear Weapons Complex: Does Anybody Care?* (Washington, D.C.: Resources for the Future, January 2000).

2. I use the term "openness" because it is in fairly common usage in regard to DOE sites and issues; "transparency," on the other hand, often appears in other public arenas. For a fine summary of what "openness" involves in the context of DOE site cleanup, see Christine H. Drew, et al., "The Hanford Openness Workshops: Fostering Open and Transparent Long-Term Decision-Making at the Department of Energy," in Thomas M. Leschine, ed., "Long-Term Management of Contaminated Sites," *Research in Social Problems and Public Policy*, Vol. 20.

3. Openness Advisory Panel, Secretary of Energy Advisory Board, *Responsible Openness: An Imperative for the Department of Energy* (Washington, D.C., August 25, 1997), p. 2. It is noteworthy that members of the Openness Panel included former senior officials in DOE's defense programs and former directors of national nuclear laboratories.

4. It is important to note that many records are protected for reasons of personal privacy. These include employment and health records for individuals.

5. Most DOE sites have searchable electronic files. Declassified and unclassified historical records are available at the department's OpenNet Web site: www.osti.gov/opennet. For places where active cleanup is complete, the Office of Legacy Management maintains and updates information at its Web site: www.lm.doe.gov.

6. The context for the release of these documents is described in Michael D'Antonio, *Atomic Harvest: Hanford and the Lethal Toll of America's Nuclear Arsenal* (New York: Crown, 1993), pp. 116 ff. The documents also formed the basis for much of the historical assessment in Michele Stenehjem Gerber, *On the Home Front: The Cold War Legacy of the Hanford Nuclear Site* (Lincoln: University of Nebraska Press, 1992, 2002). In the late 1990s, the Richland Operations Office supported a series of stakeholder workshops on openness at Hanford. For an account of the workshops, see Christine H. Drew, et al., "The Hanford Openness Workshops," op. cit.

7. USDOE, "Plutonium: The First Fifty Years: United States Plutonium Production, Acquisition, and Utilization from 1944 through 1994," available on OPENnet: www. osti.gov/opennet/forms.jsp?formurl=document/pu50yrs/pu50y.html.

8. "Classified National Security Information," Title 3, Executive Order 13292 of March 25, 2003.

9. Nick Schwellenbach, "Government Secrecy Grows Out of Control," *Corvallis Gazette-Times*, September 22, 2004.

10. These rules are not consistently applied; GAO reviewed classified documents at the Office of the Secretary of Defense and found a lack of consistency and multiple marking errors. See Government Accountability Office, *Managing Sensitive Information: DOD Can More Effectively Reduce the Risk of Classification Errors*, June 30, 2006.

11. GAO assessed the responsiveness of federal agencies to FOIA requests in a July 26, 2006, report, *Freedom of Information Act: Preliminary Analysis of Processing Trends Shows Importance of Improvement Plans*. The report's Figure 4 on page 19 indicates DOE having one of the highest rates among all agencies in fully granting requested information.

12. Dick Russel, Sanford Lewis, and Brian Keating, *Inconclusive by Design: Waste, Fraud and Abuse in Federal Environmental Health Research*. (Boston: National Toxics Campaign Fund, May 1992).

13. EEOICPA as amended, 42 USC 84, Subchapter XVI.

14. Under rules and contractual terms that extended past the end of the Cold War, the government generally indemnified its nuclear contractors against injury claims (unless in cases of negligence, which, it must be noted, often has been hard to prove). The legal costs reimbursed by DOE in fiscal years 1995–2001 amounted to more than $291 million. See Government Accountability Office letter report, "Department of Energy: Contractor Litigation Costs," March 8, 2002. According to the GAO, $625 million had been paid in lump sum compensation under EEOICPA Title B, from July 31, 2001, through January 2004. See GAO, *Energy Employees Compensation: Many Claims Have Been Processed, but Action Is Needed to Expedite Processing of Claims Requiring Radiation Exposure Estimates*, September 2004. Title B is administered by the Department of Labor and deals with radiogenic cancers.

15. Until late 2004, DOE administered Part D of EEOICPA dealing with other forms and causes of illness. The processing of claims was extremely slow; Congress moved this activity to the Department of Labor. DOE's contractual costs for granting a relatively small number of these claims were regarded as being excessive. See GAO, *Department of Energy Office of Worker Advocacy: Deficient Controls Led to Millions of Dollars in Improper and Questionable Payments to Contractors*, May 2006.

16. The Integrated Safety Management approach is set out in: DOE P 450.4, Safety Management System Policy. For information on this approach and its application at DOE sites, see: www.eh.doe.gov/ism.

17. Milton Russell, *DOE Legacy Waste Cleanup and Stewardship: Beyond the Top-to-Bottom Review*, Joint Institute for Energy and Environment, Report No. 2002-06, August 2002.

18. For an excellent summary discussion prepared by Resources for the Future in regard to trust funds, see Carl Bauer and Katherine N. Probst, *Long Term Stewardship at Contaminated Sites: Trust Funds as Mechanisms for Financing and Oversight* (Washington, D.C.: Resources for the Future, 2000).

19. Ibid., p. ix.

20. Ethan Brown, *Funding Long-term Stewardship of USDOE Weapons Sites: Tennessee's Perpetual Care Trust Fund* (Washington, D.C.: National Governors' Association Center for Best Practices, 2002).

21. See the Presidio Trust Web site: www.presidio.gov/trust.

22. John B. Price and Max S. Power, "A Washington State Perspective on Long-Term Stewardship at Hanford," paper delivered at Waste Management '02, Tucson, Arizona.

23. At the direction of Congress, the National Park Service in early 2006 began assessing four Manhattan Project sites, including Hanford, Los Alamos, Oak Ridge, and Dayton, Ohio. This study may result in proposals for National Park Service administration at one or more of these locations by the end of the decade. See: parkplanning.nps. gov/PlanProcess.cfm?parkId=422&projectId=14946.

24. An analogous situation (though minus any nuclear cleanup issues) can be found at the Salmon Ruins in San Juan County, New Mexico, where county voters approved a bond issue to build a museum and research center on privately-owned property where ancient Anasazi ruins are located. The museum and research library is dedicated to the study of past Native American cultures and operated by a nonprofit organization, the San Juan County Museum Association, which has cooperative support from the county, local organizations, and the State of New Mexico. See: www.salmonruins.com.

25. National Research Council, *Long-Term Institutional Management of U.S. Department of Energy Legacy Waste Sites* (Washington, D.C.: National Academy Press, 2000).

26. Steven M. Blush, "Train Wreck along the River of Money: An Evaluation of the Hanford Cleanup," report for the U.S. Senate Committee on Energy and Natural Resources, March 1995.

27. *The Final Report of the Federal Facilities Environmental Restoration Dialogue Committee: Consensus Precipices And Recommendations For Improving Federal Facilities Cleanup* (Washington, D.C., 1996), available at: www.epa.gov/fedfac/fferdc.htm.

28. Judith A. Bradbury, Kristi M. Branch, and Elizabeth L. Malone, *An Evaluation of DOE-EM Public Participation Programs* (Richland: Pacific Northwest National Laboratory, February 2003–PNNL-14200), especially pp. 51–56, dealing with site specific advisory boards. Unfortunately, DOE's inheritance of the Manhattan Project's original secrecy and decentralization, and its overlapping and unclear lines of accountability, often have prevented the effective utilization of citizen advisory boards.

29. This account is based on an interview with Lisa Crawford, November 28, 2006.

30. This discussion is based on an interview with Lynne Stembridge, November 22, 2006.

31. The volunteers included Tim Connor, Larry Shook, Jim Thomas, and J.R. Wilkinson. Connor and Shook were gifted investigative journalists, whereas Thomas and Wilkinson were self-described dedicated "ordinary folks." Connor, Shook, and Thomas eventually became paid staff for a time. Another group member, Todd Martin, joined the HEAL staff later; he was the one staff member with an academic training in things scientific.

32. HEAL's publications—which are models for good analysis—are not available in electronic form. Hopefully, they have been retained in collections at the Hanford information repositories. These libraries are listed at: www.hanford.gov.

33. William L. Riordon, *Plunkitt of Tammany Hall* (New York: E.P. Dutton, 1963), pp. 17–20.

Index

A

Abraham, Spencer (Sec. of Energy), 82
Agency for Toxic Substances and Disease
    Registry (ATSDR), 48, 163, 164
Alaska, 95, 96
Albuquerque, New Mexico, 8
Alliance for Nuclear Accountability
    (ANA), 57, 173, 175
aluminum clad uranium, 125–26
americium, 42
Andrus, Cecil (Idaho Gov.), 91–92, 113,
    127
Arid Lands Ecology Reserve, 73
Atomic Energy Act (1946), 34, 36, 44, 45,
    49–50
Atomic Energy Commission (AEC), 4–5,
    9, 33, 42, 46, 50, 52, 78, 89, 162
Atoms for Peace, 4

B

Batt, Phil (Idaho Gov.), 92, 113
Battelle, 61, 110
Bechtel, 139
Bodman, Samuel (Sec. of Energy), 98
Bonneville Power Administration (BPA),
    51
B-Reactor, 138, 168, 169
breeder reactor(s), 53, 62
Brookhaven National Laboratory (BNL),
    21, 112, 119
Bush, George W., administration, 8, 15,
    145, 161

C

California, 92, 95, 96, 97
canyon(s), 75, 136, 137, 139, 154
Carlsbad, New Mexico, 5, 13, 44, 56, 60,
    92, 93, 127
Carter, Jimmy, administration, 51, 53, 61
Centers for Disease Control, 48, 163, 164
cesium, 16, 17–18, 120, 130
    cesium-137, 10, 11

chemical waste/hazards, 5, 11, 31, 38, 39,
    43, 77, 81, 103, 105, 109, 122–23, 166
Chernobyl, 10, 14, 26, 31, 51, 56, 58, 110
Chicago, Illinois, 8, 104
chromium, 3, 126
Cincinnati, Ohio, 119
Clean Air Act (CAA), 35, 36, 39
Clean Water Act (CWA), 35, 36, 39
Clinton, Bill, administration, 103, 154,
    177
Clinton Engineer Works, 4
Cold War, 3–4, 7, 11, 33, 35, 50, 61, 73, 74,
    117–24 passim, 137, 141, 146, 162–63,
    165, 182
Colorado, 37, 92, 96, 171
    Gov. Romer, 92, 109
Colorado Public Service Company, 92
Columbia Generating Station, 138
Columbia River, 4, 51, 52, 69–70, 74–75,
    78–79, 119, 120, 123, 126, 140, 148,
    162–63, 170
Columbus, Ohio, 110
Commercial Vehicle Safety Alliance, 130
Committee for a SANE Nuclear Policy
    (SANE), 55, 57
Comprehensive Environmental Response,
    Compensation and Liability Act
    (CERCLA), 36, 38–41, 45, 73, 78, 79,
    103
Confederated Tribes of the Umatilla
    Indian Reservation, 85, 86
Connecticut, 96, 112
Council of Energy Resource Tribes
    (CERT), 85, 173
Council on Environmental Quality, 102
Crawford, Lisa, 174, 176–78
Critical Infrastructure Information (CII),
    36, 161

D

Deaf Smith County, Texas, 55, 99
Defense Nuclear Facilities Safety Board
    (DNFSB), 40, 120, 121
Defense Waste Processing Facility
    (DWPF), 21, 146–48

DOE Organization Act (1977), 36
downwind/down-winders, 162, 164
D-Reactor, 138, 141–42
DuPont, 8

E

East Tennessee Technology Park, 50
Eisenhower, Dwight D., administration, 4
Energy Communities Alliance (ECA), 57,
    173, 175
Energy Employees Occupational Illness
    Compensation Program Act
    (EEOICPA), 36, 164–65
Energy Northwest, 138
Energy Reorganization Act (1974), 34,
    36, 42, 46
Energy Research and Development
    Administration (ERDA), 33, 46, 52
Environmental Restoration Disposal
    Facility, 75

F

fast breeder reactor, 61
Fast Flux Test Facility (FFTF), 61–62,
    63, 139
Federal Facilities Compliance Act (1992),
    36, 38, 105, 109
Federal Facilities Task Force, 105, 175
Federal Highway Administration (FHA),
    48
Federal Railroad Administration (FRA),
    48
Feed Materials Evaluation Facility
    (FMEF), 62, 139
Fernald Residents for Environmental
    Safety and Health (FRESH), 57,
    176–78, 180–81
Fernald site (Ohio), 3, 21
    closure, 12, 47, 58–59, 60, 73, 83, 126,
        170, 172, 174, 176–78
    Dewatering Excavation Evaluation
        Program, 177
    Feed Materials Production Center,
        174
    soil/water (surface/ground), 119, 126,
        176, 177
ferrocyanide, 37, 120
Fort St. Vrain reactor, 92

Freedom of Information Act (FOIA),
    36, 161
    Critical Infrastructure Information,
        36, 161

G

Gable Mountain, 1, 68
Gardner, Booth (Washington Gov.), 24
Glenn, John (U.S. Senator), 56
Government Accountability Project
    (GAP), 56, 57, 173
Government Operations Committee, 56
granite formations (geological), 54–55
Great Miami Aquifer, 126
Gregoire, Christine, 37
grout/grouting, 12, 72

H

Hanford Education Action League
    (HEAL), 55, 57, 178–81
Hanford Natural Resource Trustee
    Council (NRTC), 79–80
Hanford Reach, 2
Hanford Reach National Monument, 74,
    140, 168
Hanford site (Washington), 21, 172,
    178–79
    100-K Area, 6
    101-SY Tank, 120
    183-H Basin, 94
    24 Command Fire, 17–18
    242-Z facility, 144
    300-Area, 18, 74
    Arid Lands Ecology Reserve, 73
    B-Reactor, 138, 168, 169
    Canister Storage Building (CSB),
        106–7
    Columbia Generating Station, 138
    cribs, 150–51
    D-Reactor, 138, 141–42
    Environmental Restoration Disposal
        Facility, 75
    Fast Flux Test Facility , 61–62, 63, 139
    Feed Materials Evaluation Facility,
        62, 139
    Gable Mountain, 1, 68
    Hanford Comprehensive Land Use
        Plan, 73

Hanford Defense Waste Environmental Impact Statement, 56
Hanford Dose Reconstruction and Thyroid Disease studies, 164
Hanford Engineer Works, 1
Hanford Environmental Dose Reconstruction Project, 6, 15, 22
Hanford Federal Facility Agreement and Consent Order, 37–38
Hanford Future Site Uses Working Group, 73–74, 80
High-Level Waste Vitrification Facility, 147
Interim Safe Storage, 64, 137–38
K-Basins, 51, 61, 139, 140
K-East Reactor, 19, 61, 138
Nagasaki, 2, 27
N-Reactor, 3, 34, 51, 55, 58, 60–61, 125, 138
Pacific Northwest National Laboratory, 6, 15, 21, 47, 61
Plutonium Finishing Plant, 119, 123, 125, 137, 139, 140, 157
PUREX Plant, 55, 58, 61, 119, 137, 139, 140, 178
Rattlesnake Mountain, 1, 73–74
reactors (chart), 138
Richland Operations Office, 15, 37, 104, 160, 182
separation plants (chart), 139
soil/water (surface/ground), 80, 106, 126, 150–51, 163, 165
tank farms, 3, 6, 12, 18, 37, 56, 63, 107–8, 119–20, 122–24, 130, 139, 142, 146–48
Trinity test, 2
Wahluke Slope, 1, 74
Waste Receiving and Processing Facility (WRAP), 159
Hanford Watch, 57
Hawaii, 95, 96
heavy metals, 11, 27
hexavalent chromium, 6
high level waste (definition), 41–43, 76–78, 105, 107, 109, 115, 143, 149
Hiroshima, 2, 20, 26, 27
Homeland Security Act (2002), 36, 161
Critical Infrastructure Information, 36, 161

Houff, William, 178
hydrogen gas, 120

I

Idaho, 37, 88, 95, 96, 108–9, 149
Gov. Andrus, 91–92, 113, 127
Gov. Batt, 92, 113
Idaho National Laboratory (INL), 3, 15, 20, 21, 40, 47, 59, 86, 124, 172
cleanup, 70, 127, 141, 149, 150, 170
INEL, 15, 91–92, 105, 108, 112
INEEL, 15
soil/water (surface/ground), 119, 151, 163
tanks, 146, 149
Idaho Settlement Agreement, 127, 149
Illinois, 94, 95, 96
Indian tribes, 6, 71, 79–86 *passim*, 91, 97–98, 103, 108, 129–33, 163, 166–69, 173
Integrated Safety Management System (ISMS), 166
Interim Safe Storage (ISS), 64, 138–39
Interior and Insular Affairs Committee, 99
iodine-129, 151
iodine-131, 10–11, 26, 162

J

Jackson, Henry M. (U.S. Senator), 51
Jim, Russell, 169
Joint Committee on Atomic Energy, 4
Joint Institute for Energy and Environment (JIEE), 21, 167

K

K-Basins, 51, 61, 139, 140
K-East Reactor, 19, 61, 138
Kennedy, John F., administration, 3, 55
Kennewick, Washington, 60
Kentucky, 94, 96
Kyshtym, Russia, 122

L

Lawrence Livermore National Laboratory (LLNL), 9, 21, 47
Lawrence, Mike, 37, 160–61

League of Women Voters, 63, 104
Leroy, Steve, 160
Locke, Gary (Washington Gov.), 90
Lodge, Edward (U.S. District Judge), 113
Long Island (New York), 112, 119, 171
Los Alamos site (New Mexico), 4, 9, 20, 40, 47, 49, 59–60, 86, 93, 105, 172
    Los Alamos National Laboratories (LANL), 16, 21, 39
    Pueblo Canyon, 12
    soil/water (surface/ground), 126
Love Canal site (New York), 81
Low Level Radioactive Waste Policy Act (1980), 36, 53, 95–97, 103
low level waste compacts, 95–97
low level waste (definition), 42–44, 77–78, 105, 115, 132
Lowe, Adam, 157, 170, 173
Lyons, Kansas, 52

M

Madison, James, 181
Magnuson, Warren (U.S. Senator), 51
Maine, 96
Manhattan Project, 2, 4–5, 8, 46, 49–50, 101, 119, 139, 162–63
    Manhattan Engineering District, 14, 36
Marshall Islands, 164
Massachusetts, 96
McDonald, Alan (U.S. District Judge), 44
McKellar, Kenneth (U.S. Senator), 4
Miamisburg, Ohio, 59, 68
Michigan, 95, 96
Model Toxics Control Act (Wash. state), 29
Montana, 95, 96
Morris, Illinois, 53
Mound Facility (Ohio), 21
    closure, 59, 68, 170
MOX fuel, 62, 63, 65–66

N

Nagasaki, 2, 27
National Academy of Sciences, 52, 121
National Conference of State Legislators (NCSL), 173
National Congress of American Indians (NCAI), 85, 173

National Environmental Policy Act (NEPA), 36, 56, 102, 145
National Governors Association (NGA), 105, 173, 175
    Federal Facilities Task Force, 105, 175
National Institute for Occupational Safety and Health (NIOSH), 48, 164
National Nuclear Security Administration (NNSA), 47, 125, 140
National Research Council (NRC), 99, 121, 152, 170
Native American tribes, 6, 71, 79–86 passim, 91, 97–98, 103, 108, 129–33, 163, 166–69, 173
Natural Resources Defense Council (NRDC), 16, 56, 57, 102, 107–9, 173
Naturally Occurring Radioactive Material (NORM), 43
Navajo Nation, 85, 86
naval propulsion/fuel, 2, 3, 46–47, 91–92, 124–25, 127
Nazi Germany, 2
Nebraska, 95, 96
neptunium, 42
Nevada, 37, 60, 92, 94, 96, 99–101, 111
    Nevada Agency for Nuclear Projects, 175
Nevada Test Site, 4, 5, 21, 25, 40, 54–55, 60, 77, 101, 103, 105, 151, 172, 178
    Area-3 Radioactive Waste Management, 131
    Sedan crater, 5
    soil/water (surface/ground), 163
    Yucca Mountain Repository, 12, 21, 42, 54, 60, 76, 86, 98–103, 106, 152, 175
New Hampshire, 96
New Jersey, 96, 112
New Mexico, 37, 92–93, 96, 106–7, 109–10, 115, 116, 128–29
    Environment Department, 129
    Richardson, Bill, 93, 107, 115, 127
New Mexico pueblos, 85, 86
New York City, 112
New York (state), 94, 96, 109, 112, 171
Nez Perce Tribe, 85, 86
Niagara Falls, New York, 81
"9/11," 160, 161, 171
NORM, 43

North Carolina, 95, 96
N-Reactor, 3, 34, 51, 55, 58, 60–61, 125, 138
Nuclear Control Institute (NCI), 57, 63, 175
Nuclear Freeze, 55
Nuclear Information and Resource Service, 175
Nuclear Regulatory Commission (NRC), 24, 34, 42–43, 45, 46, 52, 108, 121, 130, 132–33
Nuclear Waste Policy Act (NWPA), 35, 36, 41, 54, 58, 86, 97–99, 107, 109, 152
    Nuclear Waste Negotiator, 99–100
    Nuclear Waste Policy Amendments Act (1987), 99, 103
    Nuclear Waste Technical Review Board (NWTRB), 102, 121, 153
    See also, Yucca Mountain Repository
Nuclear Weapons Freeze Campaign (NWFC), 57

O

Oakland, California, 8, 111
Oak Ridge site (Tennessee), 2–4, 8, 27, 47, 49–51, 59, 73, 105, 116, 125, 168, 172
    cleanup, 70, 104, 110, 126, 128, 141, 170
    Clinton Engineer Works, 4
    East Tennessee Technology Park, 50
    Hiroshima, 2, 27
    Joint Institute for Energy and Environment, 21, 167
    Oak Ridge National Laboratory (ORNL), 21
    soil/water (surface/ground), 11, 35, 119, 120, 126, 151, 163, 170
    Y-12 plant, 4, 140
Oak Ridge, Tennessee, 49
Office of Civilian Radioactive Waste Management (OCRWM), 23–24, 47
Office of Environmental Management (EM), 7, 15, 37, 47, 117–18, 125–26, 143, 170, 175
Office of Health, Safety, and Security, 47
Office of Legacy Management (LM), 47, 84, 178
Office of Nuclear Energy, 47
Office of River Protection (ORP), 107

Office of Science, 47
Office of Secure Transportation (OST), 47
Ohio, 37, 88, 92, 96, 118, 171
    Ohio Environmental Protection Agency, 176
Openness Panel, 158–61
Oregon, 80, 95, 96, 104, 113, 116
    Hanford Natural Resource Trustee Council, 79–80
    Nuclear Safety Division, 42, 45
    nuclear waste definitions, 43, 109
    Oregon Hanford Cleanup Board, 155
Owen, Michael, 90
Owens, Bill (Governor), 90

PQ

Pacific Islanders, 26
Pacific Northwest National Laboratory (PNNL), 6, 15, 21, 47, 61
Paducah site (Kentucky), 3, 21
    gaseous diffusion plant, 119
    soil/water (surface/ground), 119
Pasco, Washington, 60
PCBs, 39
Peace Action, 57
Peña, Federico (Sec. of Energy), 161
Pennsylvania, 95, 96
Physicians for Social Responsibility (PSR), 57, 63, 104
Plunkitt, George Washington, 180
plutonium, 2, 12, 17–18, 41–43, 51, 55, 66, 75, 123–25, 141, 145, 161–62, 170, 178–79
    Pu-239, 11, 27
    soil/groundwater, 68, 101, 119, 120
    weapons useable, 3, 7, 16, 27, 53, 58, 61–63, 104, 123, 140, 143, 146
Plutonium Finishing Plant (PFP), 119, 123, 125, 137, 139, 140, 157
Plymouth ,Washington, 112–13
Portland, Oregon, 113
Portsmouth site (Ohio), 3
Presidential Ex. Order 13084 (Indian Tribal Governments), 36, 86
Presidential Ex. Order 13392 (Disclosure of Information), 36, 161
President's Council on Environmental Quality (CEQ), 121

President's Foreign Intelligence Advisory
    Board, 16
Presidio Trust, 168
Price, John, 168, 184
Probst, Katherine, 157, 170, 173
Programmatic Environmental Impact
    Statement, 102–3
Project Gnome, 5
Pueblo Canyon, 12
PUREX Plant, 55, 58, 61, 119, 137, 139,
    140, 178

R

radium, 43, 44
radon, 43, 174
Rattlesnake Mountain, 1, 73–74
Ray, Dixie Lee (Washington Gov.), 95,
    112
Reagan, Ronald, administration, 51, 54,
    55
Resource Conservation and Recovery Act
    (RCRA), 5, 35, 36, 38–41, 44, 109
Resources for the Future, 8, 57, 157
Rhode Island, 96
Richardson, Bill (Rep./Gov./Energy Sec.),
    93, 107, 115, 127
Richland, Washington, 1, 8, 17, 49, 60,
    120, 160, 168
Richland Operations Office, 15, 37, 104,
    160, 182
Rockwell, 56
Rocky Flats site (Colorado), 3, 21, 39, 40,
    49, 55, 91, 162
    closure, 23, 47, 58–59, 67–69, 72, 73,
        75, 83, 92, 109, 124, 126–28, 158,
        170
    F&WS preserve, 68, 140, 170, 172
    FBI investigation, 35, 56, 65
    soil/water (surface/ground), 119, 126
Romer, Roy (Colorado Gov.), 92, 109
Roosevelt, Franklin D., administration, 4
Russell, Milton, 88, 167
Russia/Russians, 62, 63

S

salt formations (geological), 52–53, 54
Sandia National Laboratories (SNL/NM),
    21, 86, 87
San Francisco area, 8, 111, 168, 171

Savannah River site (South Carolina), 3,
    8, 21, 27, 49, 59, 63, 125–26, 140, 172
    cleanup, 63, 70, 105, 141–42, 145
    Defense Waste Processing Facility, 21,
        146–48
    Effluent Treatment Project, 88
    F-Canyon, 137, 154
    L-Area Basin, 163
    M-Area Settling Basin, 127
    Savannah River National Laboratory
        (SRNL), 11, 21
    soil/water (surface/ground), 119, 126,
        127, 163
    tanks, 12, 63, 108, 124, 146–48
Secretary of Energy Advisory Board, 158,
    161
Sedan crater, 5
Seneca Nation, 85, 86
Shoshone-Bannock Tribes, 85, 86, 91, 108
Snake River Alliance, 57, 108
Snake River Plain Aquifer, 35, 70, 91, 119
South Carolina, 37, 92, 94, 95, 96, 108–9
Soviet Union, 3–4, 10, 13, 31, 51, 55, 56,
    58, 61, 65, 122
Spokane, Washington, 178, 179
State and Tribal Government Working
    Group (STGWG), 143, 173, 175
Stembridge, Lynne, 178–80
Strategic Arms Limitation Treaty (SALT)
    talks, 65
strontium, 16, 120
    strontium-90, 11
"Superfund," *see* CERCLA

T

Tacoma, Washington, 78, 111
technetium-99, 151
Tennessee, 37, 92, 96, 116, 168
Tennessee Valley Authority (TVA), 4,
    48, 65
Texas, 96, 97, 99
Three Mile Island, 13, 31, 51, 53, 55, 91
transmutation, 77
transportation
    Office of Secure Transportation
        (OST), 47
    Transportation External
        Coordination Working Group
        (TEC), 173

TRUPACT-II, 130
See also, U.S. Department of
Transportation
transuranic waste (definition), 41–44, 76,
105–7, 109, 143, 149, 151, 170
Tri-City Economic Development
Council, 63, 104
Trinity test, 2
tritium, 65–66, 158
Trojan power plant, 112–13, 116
TRUPACT-II, 130

UV

Udall, Morris K. (U.S. Rep., Arizona), 99
Ukraine, 10
uranium, 2–3, 16, 41, 45, 53, 55, 58, 62,
75, 124, 140, 176
highly enriched uranium (HEU),
27, 125
low enriched uranium (LEU), 27
mining/milling, 2, 27, 36, 84, 164
soil/groundwater, 119
U-235, 27, 65
U-238, 27, 65
Uranium Mill Tailing Remediation
Control Act (UMTRCA), 36
U.S. Army, 168
U.S. Army Corps of Engineers (USACE),
14, 33, 48
U.S. Attorney General, 161
U.S. Bureau of Land Management
(BLM), 48, 79
U.S. Coast Guard (USCG), 48
U.S. Department of Defense (DOD), 48,
171–72
U.S. Department of Energy (DOE or
USDOE), 46, 181
American Indian Policy, 86
Bonneville Power Administration, 51
DOE Organization Act, 36
Energy Employees Occupational
Illness Compensation Program
Act, 36, 164–65
Integrated Safety Management, 166
National Nuclear Security
Administration, 47, 125, 140
Office of Civilian Radioactive Waste
Management, 23–24, 47

Office of Environmental
Management, 7, 15, 37, 47,
117–18, 125–26, 143, 170, 175
Office of Health, Safety, and Security,
47
Office of Legacy Management, 47,
84, 178
Office of Nuclear Energy, 47
Office of River Protection, 107
Office of Science, 47
Office of Secure Transportation, 47
Programmatic Environmental Impact
Statement, 102–3
Secretary of Energy Advisory Board
Openness Panel, 158–61
State and Tribal Government
Working Group, 143, 173, 175
Transportation External
Coordination Working Group,
173
Waste Management Programmatic
EIS, 103
U.S. Department of Health and Human
Services (HHS), 48
Agency for Toxic Substances and
Disease Registry, 48, 163, 164
Centers for Disease Control, 48, 163,
164
National Institute for Occupational
Safety and Health, 48, 164
U.S. Department of Homeland Security
(DHS), 36, 48
U.S. Coast Guard, 48
U.S. Department of the Interior, 169
U.S. Department of Transportation
(DOT), 45, 48, 112–13, 130, 132
Federal Highway Administration, 48
Federal Railroad Administration , 48
U.S. Environmental Protection Agency
(EPA), 18, 24, 37–41, 44, 48, 73, 79,
82, 93, 110, 129
CERCLA/"Superfund," 36, 38–41, 45,
73, 79, 103
U.S. Federal Bureau of Investigation
(FBI), 35, 56, 65
U.S. Fish and Wildlife Service (FWS), 48,
68, 79, 140, 168, 170
U.S. Government Accountability Office
(GAO), 15, 48, 121, 145

U.S. Navy, 92; *see also*, naval propulsion/fuel
U.S. Office of Management and Budget (OMB), 48, 82–83, 161
U.S. Supreme Court, 33, 44
Utah, 88, 97, 118, 164, 178
vitrification, 12, 63, 66, 145–48, 155

W

Wahluke Slope, 1, 74
Washington (State of), 24, 37, 44, 58, 80, 88, 92, 94–96, 99, 104, 107–13
   "Don't Waste Washington" initiative, 53, 95
   Ecology Department, 37
   Gov. Gardner, 24
   Gov. Locke, 90
   Gov. Ray, 95, 112
   groundwater law, 80
   Hanford Future Site Uses Working Group, 73–74, 80
   Hanford Natural Resource Trustee Council, 79–80
   Model Toxics Control Act, 29
Waste Isolation Pilot Plant (New Mexico), 5, 13, 21, 44, 56, 60, 76, 92–93, 103–10, 118, 127–32, 149, 151, 158
   Commercial Vehicle Safety Alliance, 130
   Environmental Evaluation Group, 128
   WIPP Land Withdrawal Act (1992), 36, 92, 103

Waste Management Programmatic EIS, 103
Watkins, James (Sec. of Energy), 6, 35, 102–3
Weldon Spring site (Missouri), 3, 21, 39, 171
   closure, 58–59, 83, 126, 170, 172
West Valley site (New York), 12, 21, 53, 86, 109, 155
   West Valley Demonstration Project Act (1980), 36, 155
Western Governors Association, 99, 130
Western Shoshone Tribe, 85, 86
White Bluffs, 1
Women Against Nuclear Armament, 57, 178
WPPSS, 138
WSU Tri-Cities, 17
Wyden Watch List, 122
Wyoming, 95, 96

XYZ

xenon gas, 10
Y-12 plant, 4, 140
Yakama Indian Nation, 80, 85, 86, 108, 169
Yucca Mountain Repository (Nevada), 12, 21, 42, 54, 60, 76, 86, 98–103, 106, 152, 175; *see also*, Nuclear Waste Policy Act
zirconium clad uranium, 126

About the Author

Max S. Power earned a B.A. in political science as a National Merit Scholar at The Colorado College, a B.A. in philosophy, politics, and economics as a Rhodes Scholar at Oxford University, and M.A., M.Phil., and Ph.D. degrees in political science at Yale University where he was a Yale University Fellow and Danforth Fellow.

Following a teaching stint at the University of Victoria, British Columbia, in 1969–1972, Power served for 13 years as an intergovernmental relations specialist, regional-development project manager, land-use planning director, and policy analyst for local governments in the Puget Sound area.

In 1984–1985, he became staff director for the Washington legislature's Joint Select Committee on Science and Technology, and in 1985–1987 was principal investigator for the Washington State Institute for Public Policy at The Evergreen State College.

From 1987 to 2004, Power held various positions, including senior policy adviser, in the Nuclear Waste Program of the Washington State Department of Ecology, Olympia, where he worked on nuclear contamination and cleanup issues. In this period, he participated in the creation of the Hanford Joint Council for Employees Concerns, the Hanford Future Site Uses Working Group, and the Hanford Advisory Board. He also was involved in the Hanford Openness Workshops, and served in joint projects with the Washington League of Women Voters and other nonprofit and governmental organizations in a series of workshops regarding the disposition of surplus plutonium and the interstate transfers of nuclear waste.

Power also has been a member of DOE's State and Tribal Government Working Group, DOE's Transportation External Coordination Working Group, the High Level Radioactive Waste Committee of the Western Interstate Energy Board, and the Federal Facilities Task Force of the National Governors Association.

Max Power resides in Corvallis, Oregon. Currently he is serving as a governor-appointed public member of the Oregon Hanford Cleanup Board, and is a neutral member of the Hanford Concerns Council.